GLOBAL DESERTS OUTLOOK

Produced by
Division of Early Warning and Assessment (DEWA)
United Nations Environment Programme
P.O. Box 30552
Nairobi 00100, Kenya
Tel: (+254) 20 7623562
Fax: (+254) 20 7623943
E-mail: dewa.director@unep.org
Web: www.unep.org

Editor: Exequiel Ezcurra, San Diego Natural History Museum

Graphics and layout: Printing and Publishing Section, UNON
Printing: Scanprint, Denmark
Distribution: SMI (Distribution Services) Ltd. UK
This publication is available from Earthprint.com http://www.earthprint.com

Front cover photo: White Sands National Monument, New Mexico, Kevin Schafer / Still Pictures

Inside covers: Desert Scene, by Ali Selim. Werner Forman / Art Resource, NY

Global Deserts Outlook

Edited by Exequiel Ezcurra

Contents

Foreword

The world's deserts represent unique ecosystems which support significant plant and animal biodiversity, particularly with respect to adaptations for survival in arid conditions. Various human societies have also been established in deserts throughout history, and today deserts are an important part of the world's natural and cultural heritage. Deserts are also diverse landscapes, contrary to the common notion of vast swathes of endless sand; for example, the FAO-UNEP Land Cover Classification System has identified over seventy classes of desert land cover in Egypt alone.

Desertification – the degradation of drylands due to factors including climatic variations and human activities – is among the most serious environmental challenges facing the world today. Deserts cover a total of over 19 million square kilometres, representing almost 15 per cent of the terrestrial surface of the planet, and are currently home to some 144 million people. At the same time, poverty affects many of the people living in deserts. Moreover, desertification is a truly global problem, affecting areas and populations outside of drylands. Dust from the Gobi and Sahara deserts has, for instance, been linked to respiratory problems in North America and has affected coral reefs in the Caribbean Sea.

It is against this backdrop that the UN General Assembly has declared 2006 to be the International Year of Deserts and Desertification (IYDD). The international community is deeply concerned by desertification, its implications for the achievement of the Millennium Development Goals, and the need to raise further awareness of the issues.

UNEP, as part of its global environmental assessment programme, and as a contribution to IYDD, has undertaken this global assessment of deserts. The *Global Deserts Outlook* represents the first thematic assessment report in UNEP's Global Environment Outlook series. The special focus of this report, in the context of IYDD, will help to raise global public awareness of the state and development potential of the world's deserts. The report draws on various studies and assessments of dryland ecosystems that have yielded valuable new insights into the issues of deserts and desertification, although significant gaps in terms of data and methodologies remain.

The *Global Deserts Outlook* provides a balanced picture of deserts. It shows that they are more than landscapes which are the end result of the process of desertification. The report urges policy-makers to consider the development potential of deserts, and their conservation needs: what are the most appropriate and sustainable livelihoods for people living in desert areas? Although deserts do not have much water, they do have other valuable natural resources that benefit people, such as biological and cultural diversity, and minerals. They also have the potential to attract tourists and generate solar power. The scientific knowledge and engineering skills needed to generate sustainable incomes from desert resources already exist; appropriate actions and equitable sharing of the proceeds need to be determined.

The *Global Deserts Outlook* is a stimulating and informative resource for all those concerned with deserts and desertification, and the sustainable development of dryland environments.

Shafqat Kakakhel
United Nations Assistant Secretary-General and Deputy Executive Director
United Nations Environment Programme

Executive Summary

DESERTS HARBOUR RICH ECOSYSTEMS

Deserts cut cross our planet along two fringes parallel to the equator, at 25–35° latitude in both the northern and southern hemispheres. The Desert Biome can be defined climatologically as the sum of all the arid and hyper-arid areas of globe; biologically, as the ecoregions that contain plants and animals adapted for survival in arid environments, and, physically, as large contiguous areas with ample extensions of bare soil and low vegetation cover. A map produced by overlaying areas under these three criteria shows a composite definition of the world's deserts, occupying almost one-quarter of the earth's land surface, some 33.7 million square kilometres.

Deserts landscapes are diverse; some are found on a flat shield of ancient crystalline rocks hardened over many millions of years, yielding flat deserts of rock and sand such as the Sahara, while others are the folded product of more recent tectonic movements, and have evolved into crumpled landscapes of rocky mountains emerging from lowland sedimentary plains, as in Central Asia or North America.

Over the last two million years — the Pleistocene period — climatic variations of the earth have transformed the world's deserts, forcing them to shrink during cold glacial periods and expand during the hot interglacials, leading finally to the current warming and aridization trend of the last 5 000 years, from the mid-Holocene to date. Some of the Ice-Age species still survive in arid mountain ranges, or desert "sky-islands", as rare relictual organisms.

Most large deserts are found away from the coasts, in areas where moisture from the oceans rarely reaches. Some deserts, however, are located on the west coasts of continents, such as the Namib in Africa, or the Atacama in Chile, forming coastal fog-deserts whose aridity is the result of cold oceanic currents.

The deserts of the world occur in six global bio-geographical realms:

- The **Afrotropic deserts** are found in the sub-Saharan part of Africa, and in the southern fringe of the Arabian peninsula. Their mean population density is 21 persons per square kilometre, and their human footprint (that is, pressures on the environment resulting from human activities) is relatively high, especially in the Horn of Africa and Madagascar.
- The **Australasian deserts** comprise a series of lowland arid ecoregions in the Australian heartland, covering in total some 3.6 million square kilometres, of which some 9 per cent is under some degree of environmental protection. Hardly inhabited at all, their mean population density is less than 1 person per square kilometre, and show, by far, the lowest human footprint among the global deserts.
- The **Indo-Malay** region harbours only two hot lowland deserts — the Indus Valley and the Thar — covering in total 0.26 million square kilometres, of which some 20 per cent receives some level of legal environmental protection. With a mean density of 151 persons per square kilometre, these are the deserts with the most intense human use in the world.
- The **Nearctic deserts** cover 1.7 million square kilometres in North America, of which 19 per cent is under some level of legal protection. Because of the growth of large urban conglomerates such as Phoenix in the United States, their mean population density is high (44 persons per square kiometre) and their mean human footprint (21) is the second highest of the world's deserts, especially in the Sonoran and Chihuahuan deserts.
- The **Neotropic deserts** in South America cover 1.1 million square kilometres, of which only 6 per cent receives legal protection. Their mean population density is 18 persons per square kiometre, and their mean human footprint (16) is lower than in their North American

counterparts, with most pressure concentrating in the Sechura Desert in the coasts of Peru.

- By far, the **Paleartic realm** concentrates the largest set of deserts in the world, covering a remarkable 16 million square kilometres that total 63 per cent of all deserts on the planet. Their population density is 16 persons per square kilometre, and their mean human footprint (15) is the second lowest on the planet, possibly because of their sheer inaccessibility and extreme aridity. The Sahara, an immense shield-desert, occupies 4.6 million square kilometres, or 10 per cent of the African continent. In sharp contrast with the flat Sahara and Arabian deserts, the deserts of Central Asia present folded mountains with high landscape heterogeneity and enclosed basins, some of which contain large lakes such as the Caspian and Aral Seas.

With summer ground surface temperatures of near 80°C, and only enjoying very ephemeral pulses of rain, species in deserts have evolved remarkable adaptations to harsh conditions, ranging from plants adapted to the fast use of ephemerally-abundant water or to extraordinarily efficient use of scarce water, to behavioural, anatomical, and physiological adaptations in animals. Some species from different deserts show striking resemblances in their appearances despite their differences in phylogenetic origins and biogeographic histories, a phenomenon known as convergent evolution. As a survival strategy, many desert species have symbiotic interactions and cooperate with each other through pollination, fruit dispersal, or by providing protective shade.

True deserts are not the final stage of a process of desertification; they are unique, highly-adapted natural ecosystems, both providing life-supporting services on the planet and supporting human populations in much the same ways as in other ecosystems.

DESERTS ARE THE HOME OF DIVERSE CULTURES AND LIVELIHOODS

Deserts are home to many human populations of the world. Currently around 500 million people live in deserts and desert margins, totalling 8 per cent of the global population. Among the greatest contributions of desert cultures to the world are the three "religions of the Book", Judaism, Christianity and Islam, which have had tremendous impact far beyond their areas of origin.

Humans have learnt to survive in deserts, compensating for their poor morphological and physiological adaptations to desert climates with a panoply of behavioural, cultural and technological adaptations to the dry environments. Traditionally, desert livelihoods were of three types — hunter-gatherers, pastoralists, and farmers. Hunter-gatherer tribes, such as the Topnaar of the Namib, are known for their in-depth knowledge of local food plants and wild animal species. Pastoralism, on the other hand, makes use of domesticated animals, such as camels or goats, to produce products such as milk, leather, and meat. Desert agriculture occurs mostly around oases and desert rivers, which often provide silt and nutrients through flooding cycles.

These ways of life, however, are changing rapidly, from hunter-gatherers to cattle ranchers, and from nomadism and transhumance to tourist-targeted activities. Irreversible damages have been caused in previously good agricultural grounds in deserts by large-scale modern developments, such as dam constructions for water and energy supplies. In recent times, extraction of minerals, use of the vast open spaces for military facilities, energy-intensive urban developments, and tourism, have increasingly changed the ways of life for some desert populations.

Resource use and management in deserts for these developments focuses and depends heavily on water and energy, two key resources. Recent increases in the pace of desert urbanization are the result of the relocation of expansive land developments, mining and power engineering, the growth of transport infrastructure, and improvements in water extraction and supply technologies. The high, or even complete, dependency of large desert cities on imported resources has become economically feasible as they generate sufficient income from their economic activities.

Due to the extremely slow rate of biological activity in deserts, these ecosystems take decades, if not centuries, to recover from even slight damage, such as the tracks left behind by an off-road vehicle on a lichen-covered hill. Moreover, because traditional livelihoods in deserts require large areas, they are particularly vulnerable to political and environmental changes. A good example of this is how the lives of nomadic herders in the Gobi floundered under the changes from Mongolia's transition from a socialist system to a market-driven economy.

DESERTS PLAY AN IMPORTANT ROLE IN THE GLOBAL ENVIRONMENT AND ECONOMY

Deserts interact strongly with the rest of our planet. Global-scale climate change during the 1976–2000 period has shown increased temperatures in nine out of twelve deserts studied. Average projected changes for 2071–2100 show a temperature increase of between one and seven degrees Celsius in all world deserts. Rainfall, on the other hand, could increase or decrease with climate change: while the Gobi Desert in China will most likely receive more rain, the Sahara and Great Basin deserts could become drier. In general, a warmer planet will bring more rainy pulses to winter-rain deserts and more drought pulses to summer-rain deserts. Large desert rivers originate mostly outside deserts, and many could face declining water flow from climate change.

These changes will, undoubtedly, impact the ecology of deserts. For example, nearly half of the bird, mammal and butterfly species in the Chihuahuan Desert are expected to be replaced by other species by 2055. Annual grasses that are prone to wildfire are likely to extend their coverage in some deserts, invading native scrubs and increasing the risk of soil erosion.

Deserts also have strong linkages to non-desert environments. Decreased rainfall in some deserts as a result of climate change will represent increased emissions of cross-boundary dust storms with, literally, far-reaching consequences. Most dust particles in the global atmosphere originate from the deserts of northern Africa (50–70 per cent) and Asia (10–25 per cent). Nutrients carried by desert dust, such as phosphorus and silicon, enhance growth in oceanic phytoplankton by increasing the productivity of some marine ecosystems, and also of nutrient-poor tropical soils, as observed from Saharan dust deposited in the Amazon basin. Desert-generated dust also reduces visibility, interfering with ground and air traffic away from deserts and increases the incidence of respiratory illnesses.

Deserts provide migratory corridors for many species. Non-desert birds on cross-desert migration across the Sahara compete increasingly with the human population of the region for rare oases that cover only two per cent of the area. The desert locust (*Schistocera gregaria*) is normally found in 25 countries of the Sahel and the Arabian Peninsula, but during epidemic outbreaks can spread over up to 65 countries, consuming 100 000 tonnes of vegetation a day, from India to Morocco, and even crossing the Atlantic to the Caribbean and Venezuela.

Deserts have provided trade corridors from times immemorial through which goods and cultures travelled. Water-soluble salts, such as gypsum, borates, table salt, sodium and potassium nitrates have been historically a product of deserts. Evaporite minerals, such as soda, boron, and nitrates, are common in deserts and are not found in other ecosystems. A sizeable share (30–60 per cent) of other minerals and fossil energy used globally is exported from deserts, including bauxite, copper, diamonds, gold, phosphate rock, iron ore, uranium ore, oil, and natural gas.

Because of their warm climate, deserts also export agricultural products, produced under irrigation, to non-desert areas. Agriculture and horticulture are already profitable in many deserts, as in Israel and Tunisia, and have great further potential. A new non-conventional desert export is derived from aquaculture, which paradoxically, can be more efficient in water use than desert plants, and can take advantage of the deserts' mild winter temperatures and low cost of land. Biologically-derived valuable chemicals, produced by micro-algae as well as medicinal plants, are also manufactured in deserts, capitalizing on their high year-round solar radiation, and exported to global markets. Besides the ongoing export of wild plant products from deserts to non-deserts, there is a pharmaceutical potential in desert plants which is yet to be tapped.

The growth of desert cities, clearly evidenced in industrial countries in the mid-twentieth century, has attracted the migration of non-desert people into desert habitats, drawn by new employment opportunities and the availability of cheap housing. In recent years, the influx of tourists to deserts, seeking the dry and sunny climate, has encouraged migration to deserts as well. Finally, in developing countries, specifically in Sub-Saharan Africa, periodic droughts in non-desert drylands draw thousands of rural migrants and nomads to adjacent desert cities in search of food and employment.

Research carried out in deserts has enriched the knowledge of the history of our universe and planet, and of life on earth. Deserts attract scientists of every discipline, ranging from testing grounds for planetary exploration equipment, to research on meteorites (well-preserved due to the slow rate of desert rock weathering), to astronomical observations, and archaeological and geomorphologic studies. Many areas of research benefit from the desert's clean atmosphere, low human disturbance, dry climate, sparse vegetation cover, minimal cloud cover, and thin soils — features that contribute to good preservation conditions and high detectability of scientifically-relevant objects and phenomena.

Our understanding of global processes, the development of much of our modern research, our ability to cope with global environmental change, and the preservation of much of our global heritage depend to a large extent on the way we manage and preserve the world's deserts. What happens in deserts affects every one of us.

DESERTS PRESENT DEEP CHALLENGES FOR SUSTAINABLE DEVELOPMENT, BUT ALSO GREAT OPPORTUNITIES

Aside from the direct effects of reduced vegetation cover from overgrazing and deforestation, the problem of human-induced land degradation and desertification does not appear to be as serious an issue in most true deserts as it is in many semi-arid and sub-humid regions. Deserts are less susceptible to land degradation, firstly because their biological productivity is very low, and secondly because vast desert areas are almost devoid of human interference and are thus safe

from human impact. When the problem is present, it tends to concentrate on the deserts' edge or on the more humid parts inside the biome, such as oases, and desert mountain sky-islands.

In these more vulnerable portions of the global deserts, however, impacts can be significant. Removal of vegetation cover, especially due to grazing, increases soil loss. Disturbance to the fragile desert surface, by military and recreational activities, leaves long-lasting damages. Mining activities and the remnants of these have contaminated freshwater bodies with high concentrations of heavy metals and chemical substances, as seen in parts of Argentina and Chile. Oil extraction causes air pollution, spills and chronic leakages that affect both surface and subsurface organisms. Irrigated portions of deserts in China, India, and Pakistan face declining yields due to increasing salinity. In China, deterioration of the plant cover in the headwaters region of the Yangtze River has created major flooding problems downstream and massive water erosion in the Loess Plateau. While biodiversity hotspots — the biologically-richest and most endangered terrestrial ecoregions — occupy 12 per cent of deserts, almost exactly the same proportion as for hotspots globally, the proportion of the desert biome with IUCN protected area status is much less (5.5 per cent) than the same figure for all ecoregions (9.9 per cent).

People have responded to these problems by developing and implementing actions at the regional and national levels. For example, in many countries in North Africa, as well as Yemen, there is a wealth of traditional knowledge on soil and water conservation in deserts through sustainable land management practices, including the retention of suspended sediments in terraces. In an effort to make better use of investments in water-control structures, a series of protective measures have been implemented in watersheds in Tunisia and Morocco. The application of newer technologies and practices for improved fallow periods, micro-basins, windbreaks, and soil bunds has gained global momentum in light of participatory approaches to soil conservation. Since the introduction of its National Soil Conservation Program in 1983, Australia has substantially expanded and improved its soil and water conservation technologies on private and public lands.

At the international level, several assessment efforts have included the desert ecoregions; among them, the Global Assessment of the Status of Human-Induced Soil Degradation (GLASOD) conducted by the International Soil Reference Information Centre in 1988; UNEP's World Atlas of Desertification published by UNEP in 1992 and 1997; the chapter on drylands in the recent Millennium Ecosystem Assessment; and the currently on-going LADA (Land Degradation Assessment in Drylands) that started in 2006, under the auspices of several United Nations agencies. The Ramsar Convention has played a strategic role in the protection of oases and other desert wetlands. In 1994, UNCCD (the UN Convention to Combat Desertification in those Countries Experiencing Serious Drought and/or Desertification, particularly in Africa) was adopted by the international community, and 191 countries worldwide have signed or ratified the Convention so far. However, the Convention is mostly oriented towards sub-humid, semi-arid, and arid ecosystems, that is, the desert edges, and excludes the hyper-arid deserts of the world. Currently, there is no global or regional response strategy focused exclusively on deserts.

There are several forces behind environmental changes in deserts, which are also challenges to future development: changes in population dynamics will mainly affect rural desert communities along the great desert rivers. Large population increases are expected in resource-intensive populations of deserts in the United States and in the United Arab Emirates. These population changes will affect the quantities of water and energy consumed and waste produced in the desert biome. Inward investment was the strongest driver of change in deserts in the recent past; most went to the extraction of oil, gas, and minerals. Developments for nuclear weapons testing, nuclear waste, space flight, parking lots for unused aeroplanes, and other activities that have treated deserts as barren wastelands, all affect the desert environment. Tourism, another driver of change, brings nine million visitors to Morocco and Tunisia every year; there was a three-fold increase

in tourism in Egypt in 2005, and Dubai claims to be the world's fastest-growing tourist destination.

Global climate change and its impact on water regimes is already a driver of change in deserts. While rising energy prices will bring higher incomes to some oil-producing desert countries, others without this resource will suffer, as the costs of energy and water are closely correlated in deserts. Security issues from northern Africa to Iran have made deserts less accessible and have changed environmental and socio-economic conditions in these regions. Environmental problems caused by past, non-sustainable development pose enormous challenges. By far the best-known case has been that of the Aral Sea basin where the existing recovery programme will only save part of the former sea, and reduce only a proportion of the dust that the now-dry basin emits.

While building more dams and drilling for more groundwater are still tempting to policy-makers, the water in rivers that cross deserts is already thoroughly utilized, if not over-used. Groundwater, often extracted in excess of meagre recharge, rates currently provides 60–100 per cent of freshwater needs in most deserts lacking a large river. Given the escalating water crisis in many deserts, better water-use policy is urgent. Water supply can only be improved by combining new technologies with traditional water-efficient management. Useful technologies that can play an important role in future water supply include: drip irrigation and micro-sprinklers; desalination of brackish water, rather than saline water, to reduce the cost per cubic metre of treated water; fog harvesting in coastal deserts; and small sediment-holding dams and terraces.

Tourism is another opportunity for development, as long as the risks and dangers associated with it, such as volatility in the face of political conditions, competition for water and other resources, damaged beauty and biological value, temptation for street and organized-crime, social inequity, and litter, are recognized explicitly in policy. Deserts have much to offer for ecotourism, the fastest growing sector of the tourism market, although there are concerns that the label may be used to cover activities that damage ecosystems, such as off-road motoring.

Only a very small fraction of the solar power potential in deserts has been harnessed, and with the decline in the production of fossil fuel as well as technological improvements, solar sources might supply a significant portion of global energy by 2050. Wind and solar energy installations can make use of the cheap space, large inputs of solar energy, availability of some windy sites, and the absence of objectors in deserts. However, lengthy power connections required from remote desert locations are a disadvantage in both solar and wind energy production in deserts.

DESERTS WILL CONFRONT GROWING PRESSURES IN COMING DECADES

The impacts of changes in precipitation and temperature patterns due to global climate change will be highly variable from one region to the next, but they are likely to be felt the hardest in desert margins and in desert montane areas, as these are where the principal arid rangelands are located. Because deserts are driven more by climatic pulses than by average conditions, even moderate changes in precipitation and temperature may create severe impacts by shifting the intensity and frequency of extreme periods, and subsequently creating catastrophic effects on plants, animals, and human livelihoods.

Climate change is expected to affect less the total amount of available water, and more the overall water regime and the timing of water availability in deserts. Deserts and desert margins are particularly vulnerable to soil moisture deficits resulting from droughts, which have increased in severity in recent decades and are projected to become even more intense and frequent in the future. Conversely, flood events are expected to be fewer but more intense, in which case less moisture would infiltrate into soils, and run-off and eroded sediment would concentrate in depressions, reinforcing the patchiness of desert ecosystems.

Deserts fed by melting snow or ice, such as the deserts of Central Asia and the Andean foothills, will be particularly vulnerable to a changing climate. As the volume of snowpack diminishes, river regimes will change from glacial to pluvial and, as a result, total run-off is expected to increase

temporarily and then to decline. Peak discharges will shift from the summer months, when the demand is highest, to the spring and winter, with potentially severe implications for local agriculture. Growing populations in deserts and accompanying aspirations for improved standards of living, will very likely increase water demand in expanding urban areas. The deterioration of both surface and groundwater resources by agrochemicals, mostly pesticides and fertilizers used in irrigated agriculture, and increasing salinity of return flow, are likely to continue into the future. Seawater intrusion into groundwater caused by sea level rises resulting from global warming may further deteriorate the quality of underground aquifers. Desert margins, oases, and irrigated lands within deserts have a higher biological potential and are subject to increasing population pressure, and thus tend to constitute potential hotspots of degradation. Land use will continue to intensify in the desert margins while expansion of croplands into deserts will be limited, except where fuelled by irrigation. Grazing by livestock and cutting of firewood will continue to increase inside deserts, but mostly concentrated in montane areas and on the desert margins.

A decline in the rate of expansion of irrigated areas is expected in the next decades, together with increased investments in drainage to fight salinization. This would still not be enough to stop the advance of this serious problem in potential degradation hotspots including the Nile delta, the Indus, Tigris and Euphrates, and northern Mexico. A considerable amount of unsustainable irrigated land will go out of production as aquifer exhaustion progresses, and new opportunities for rehabilitation of degraded lands and sustainable pasture management systems will emerge.

Piecemeal development of infrastructure, such as road networks, will occur more in desert sky-islands and, again, in desert margins. Desert wilderness areas (any area located more than five kilometres away from any infrastructure) are expected to decline from 59 per cent of the total desert area in 2005 to a low 31 per cent by 2050, a decline of 0.8 per cent per year on average. Species such as desert bighorn sheep (*Ovis canadensis*), Asian houbara bustards (*Chlamydotis macqueenii*) and desert tortoise (*Gopherus agassizii*), that are sensitive to fragmentation of habitat or poaching, induced by increased access to the areas previously not accessible to people, will be affected significantly by this change. Relatively pristine natural rangelands inside deserts may decline by 1.9 per cent annually, and wetlands at an even higher pace, under pressure from irrigation and agricultural expansion. At greatest risk are the few patches of dry woodlands associated with desert montane habitats, which may decline by up to 3.5 per cent per year.

Currently, the desert biome holds on average an abundance of original species of 68 per cent, but the rate of biodiversity loss in deserts may double in the coming decades. A decline in original species to a mean of 62.8 per cent by 2030 and 58.3 per cent by 2050 is expected, as a result of the new pressures and impacts brought forward by agriculture and human land use (41 per cent of the loss), fragmentation associated with infrastructure (40 per cent), and climate change (6 per cent in 2000 and 14 per cent by 2050).

VIABLE OPTIONS EXIST FOR SUSTAINABLE DESERT DEVELOPMENT

Improved resource management for desert ecosystems. The extreme variability of desert ecosystems tends toward boom-and-bust cycles rather than a steady flow of environmental goods and services. Deserts, therefore, require policies that support dynamic responses to the variable and unpredictable desert environment. Mitigating the "bust" part of the cycle is an important component of the sustainable management of desert ecosystems, including not only emergency support during drought crises, but also proactive management to increase human and societal resilience, by creating diversified rural income opportunities that can sustain rural livelihoods during times of stress.

Making use of modern technology. Traditional wisdom on coping with drought, complemented by cutting-edge science and information technology holds great potential for sustainable desert resource management. Technical knowledge and reliable forecasts alone are insufficient, but need to be implemented to the benefit of the local

people. Climate change adaptation planning must therefore include the identification of vulnerable population groups and the exploration of effective and affordable livelihood strategies during times of climatic stress. Perhaps most importantly, management systems are needed that have the will and capacity to act on the most likely risk scenarios.

Renewable energy from the desert. Continuously high solar radiation makes deserts ideal locations for solar cell installations, the potential reach of which is not limited to deserts. Apart from technological feasibility, the adoption of solar energy as an alternative to fossil fuels depends on the global as well as national policy environments and concrete implementation strategies. Possible incentives to encourage the shift towards renewable energy sources include taxes on pollution-generating burning of fossil fuels, while providing loans and grants for the use of solar and other renewable energy resources.

"Soft path" for water development. Deserts, as the first environments confronted with water shortages and forced to rethink water use priorities, should be among the forerunners in developing and testing innovative, efficient, and globally-relevant, water-use technologies and policies. The "soft path" approach to water should focuses on water-use efficiency and on lowering demand, supported by economic and institutional instruments, rather than on further attempts to increase water supply. In many desert regions, water prices currently do not reflect the true cost and value of water.

A strategy to discourage wasteful water consumption, which at the same time contributes to more equitable access to water, is to support low-income and low-volume users with transparent subsidies, financed by excessive water consumers. Raising public awareness about the need to conserve water is particularly important for new migrants into deserts who have not developed a "sense of place", such as those moving into the desert cities of the U.S. Southwest.

Small-scale decentralized water supply facilities and the involvement of communities in the decision-making process about water management, allocation, and use ensure more equitable access to water and potentially lower environmental impacts than the massive centrally-planned water schemes of the 20th century. Promotion of high value-added uses of water is critical to improve water-use efficiency. For example, the high-tech industrial sector enhances the value of each cubic metre of water used many times more than the agricultural sector. Within the agricultural sector, one possibility to improve water efficiency is to restrict irrigated agriculture in deserts to high-value crops (for example, dates), intensive greenhouse farming or aquaculture, while lower-value crops such as maize can be imported from more humid regions.

A NEW VISION FOR DESERT DEVELOPMENT

Whether global deserts will follow a path of intensive development, industrial-scale agriculture projects, and mega-cities attracting massive immigration at the expense of long-term sustainability, or an alternative path of sustainable development, spurred by a "sense of place" that is sensitive to the uniqueness of the desert environment and its traditional cultures, is going to be determined by largely our common visions and collective actions taken to fulfil them.

Current desert development and conservation seems to suffer from a lack of vision and coordinated programmes. Development schemes, such as programmes for irrigated agriculture or mass tourism, tend to spring up haphazardly with few attempts to coordinate them or to plan for their long-term sustainability. Immigration to the desert is often random and opportunity-driven, and new settlements sprawl over valuable landscapes and create problems for water supply and waste management. Without proper planning and a vision of sustainability, traditional lifestyles may wither and indigenous knowledge may become lost, victims of short-term, ephemeral economic projects.

A continuation of the energy- and water-intensive development model, and a non-renewable model in which water with subsidized costs is used for low-value purposes, will not be viable, as they lead to even more severe resource depletion and degradation. On the other extreme, increased isolationism with exclusive reliance on traditional knowledge runs the risk of losing access to new sustainable technologies and might lead to diminished opportunities for younger generations, and eventually, to reduced livelihood and economic development options.

A new, more balanced vision is needed, where deserts and their inhabitants are valued by both governments and civil society; where sustainability and the well-being of desert people are given the highest priority; where desert development is guided by a long planning horizon and based on an acute understanding of the limitations and potential of these very unique environments; where market forces are harnessed to promote desert-compatible development such as low impact services or high-technology development; where traditional livelihoods are given the opportunity to survive with dignity; and where wetlands, oases, desert mountains and other fragile environments at risk are protected.

Decisions can and should be made not to change the desert, but to live with it and preserve its resources for the future. The active participation of community groups should include taking charge of their own development, planning for risks, and adapting to changing conditions while preserving their deep connections to these remarkable landscapes. The challenge remains to harness not only local, but also global policy mechanisms and market incentives to develop a viable future for deserts, where both environmental conservation and economic development are achieved.

Reader's Guide

This report is an effort to summarize and assess what we know about the challenges of the desert world. The report is organized into six sections: Chapter 1 describes the natural history and evolution of the world's deserts from a strictly ecological perspective. It shows how deserts function at a global scale, discusses the reasons of their unique richness in rare and endemic species, and analyzes their vulnerability to extinction and environmental degradation.

Chapter 2 discusses the interaction between people and deserts, both from the point of view of traditional livelihoods and from the perspective of modern desert development. It shows how people have lived in deserts for millennia and discusses the current challenges faced by these cultures for their survival. It also analyzes how people now increasingly live in desert cities or enjoy deserts temporarily for tourism or recreation.

Chapter 3 investigates the role of deserts in the planet, analyzing the linkages and interactions between deserts and non-desert environments. In an integrative fashion, it discusses global-scale connections in global climate and climate change, the influence of dust generated in deserts on the global environment, the dynamics of rivers that originate outside deserts but flow through them, the export of minerals and fossil energy from deserts to non-desert economies, the role of deserts in human transport and wildlife migration, and, finally, the role of desert research in our knowledge of the planet, of life on earth, and of peoples and their cultures.

Chapter 4 presents a general description of the environmental status of the world deserts, describing their extent, location, uniqueness, vulnerability, status of their natural resources and biodiversity, and intensity of the human footprint in their ecosystems. It analyzes the trends in land use and land degradation, and discusses the impact of land degradation in desert communities and the responses to this problem that have arisen within the international community, and also at regional and national levels.

Chapter 5 discusses the challenges confronted by deserts for their survival and development, and the opportunities that exist for sustainable development in the upcoming decades. To approach the issue, the chapter analyzes the forces that drive change in deserts, such as population, investment, or climate change, then discusses the options that exist for sustainable development with respect to problems such as water consumption and land use, and, finally, examines the opportunities that exist for the conservation and sustainable use of soil, water, endangered habitats, and biodiversity.

Finally, Chapter 6 summarizes and wraps-up the whole book with an analysis of the global outlook for deserts and the options for action. Based on a detailed analysis of trends in population, resources, and climate, the chapter explores future scenarios for water, biodiversity, and land degradation. The chapter closes with a discussion on the policy options that could lead towards sustainable management of desert resources and the enhancement of human well-being in deserts.

The report also contains two detailed numerical appendices, compiled from various sources, listing all the desert ecoregions of the world within the six biogeographic realms where deserts occur. A summary of the main traits of each desert ecoregion is presented in these appendices, including desert type, area of the whole ecoregion, area converted and under protection, species richness for plants and vertebrates, richness of endemic and threatened vertebrates, human population density, and intensity of the "human footprint" on each particular ecoregion.

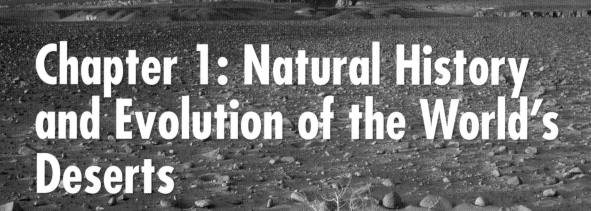

Chapter 1: Natural History and Evolution of the World's Deserts

Lead author: Exequiel Ezcurra
Contributing authors: Eric Mellink, Elisabet Wehncke, Charlotte González,
Scott Morrison, Andrew Warren, David Dent and Paul Driessen

The Desert Biome: A Global Perspective

Looking at a satellite image of the whole earth it is easy to spot a series of conspicuous ochre, vegetation-barren areas that run parallel to the equator, in both the northern and southern hemispheres, along two East-West fringes at 25-35° latitude (Figure 1.1). They are the mid-latitude deserts of the world, lying some 2 000-4 000 km away from the equatorial rainforests. In the northern hemisphere, the succession of mid-latitude subtropical deserts is formed by (1) the Mojave, Sonoran, and Chihuahuan Deserts in North America, (2) the Sahara's immense swathe in Northern Africa and the Somali-Ethiopian deserts in the Horn of Africa, and (3) the deserts of Asia, including the Arabian, Mesopotamian, Persian, and Thar deserts that stretch from West Asia into Pakistan and India, as well as the Central Asian deserts in Uzbekistan, Turkmenistan, and the Taklimakan and Gobi deserts in China and Mongolia. In the southern hemisphere, the chain is formed by (1) the Atacama, Puna, and Monte Deserts in South America, (2) the Namib and the Karoo in southern Africa, and (3) the vast expanse of the Australian deserts (Allan and others 1993, McGinnies and others 1977, Pipes 1998, Ricciuti 1996).

There are many criteria to define a desert but perhaps the most important one is aridity — the lack of water as the main factor limiting biological processes. One of the most common approaches to measure aridity is through an estimator called the Aridity Index, which is simply the ratio between mean annual precipitation (P) and mean annual potential evapotranspiration (PET, the amount of water that would be lost from water-saturated soil by plant transpiration and direct evaporation from the ground; Thornthwaite 1948). Arid and hyperarid regions have a P/PET ratio of less than 0.20; that is, rainfall supplies less than 20 per cent of the amount of water needed to support optimum plant growth (UNEP 1997, FAO 2004). Aridity is highest in the Saharan and Chilean-Peruvian deserts, followed by the Arabian, East African, Gobi, Australian, and South African Deserts, and it is generally lower in the Thar and North American deserts. Although the aridity indices vary in the different deserts in the world, all of them fall within the arid and hyperarid categories (Table 1.1).

Figure 1.1: Vegetation - barren areas of the world

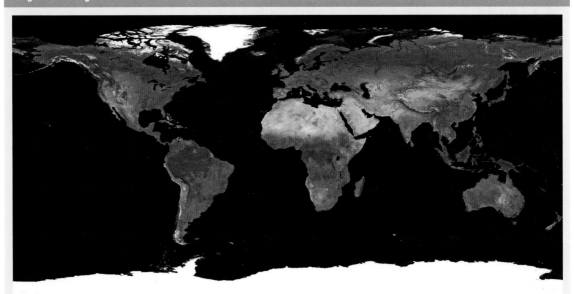

The vegetation-barren areas of the desert biome are clearly discernible in this satellite image of the earth, both north and south of the equator.

Source: NASA 2004

Table 1.1: Classification of hyperarid and arid regions of the world

The hyperarid and arid regions of the world — defined as those areas with an aridity index (*P/PET*) lower than 0.20 — cover in total some 36.2 million sq. km, and occupy almost 20 per cent of the terrestrial surface of the planet. Potential evapotranspiration (*PET*) is calculated from Thornthwaite's (1948) equations as a function of mean monthly temperatures and mean monthly number of daylight hours, while precipitation (*P*) is measured directly from weather stations.

Classification	Aridity Index (*P/PET*)	Area (km^2 × 10^6)	Area (%) of world total
Hyperarid	< 0.05	10.0	7.5
Arid	0.05–0.20	16.2	12.1

Source: UNEP 1997

Thus, the global map of arid and hyperarid regions can be used as a good approximation to the boundaries of the Desert Biome (Figure 1.2a).

A bio-ecological criterion can also be used to map the world's deserts, by lumping together all the ecoregions of the world that harbour desert vegetation (identified by the xerophilous life-forms and the general desert-adapted physiognomy of the dominant plants). The resulting set of biologically desert-like ecosystems, modified from Olson and others (2001), provides a second approximation to the Desert Biome (Figure 1.2b; see also Appendices 1 and 2).

A third criterion can be derived from AVHRR-satellite images of the world. Using a land-cover index (NDVI, or Normalized Difference Vegetation Index) the earth has been classified into different land-cover categories (GLOBIO 2005, USGS 2005). The global map of deserts and semideserts, defined as large uniform regions with extremely low vegetation cover, may be used as an alternative approximation to the Desert Biome (Figure 1.2c).

Although each approach may have its own sources of error and the three differ in their definition of what is a desert, it is surprising how the three alternative maps coincide (Figure 1.3). The Desert Biome, in short, is formed by a set of geographic regions characterized by (a) extremely high aridity, (b) a large proportion of bare soil, and (c) plants and animals showing well-defined adaptations to survive in extremely dry environments. A desert, then, is a region with very little vegetation cover and large surfaces of exposed bare soil, where average annual rainfall is less than 20 percent of the amount needed to support optimum plant growth, and where plants and animals show clear adaptations for survival during long droughts.

LATITUDINAL DESERT BELTS

Deserts occur in specific latitudes (25–35° north and south of the equator) because of the general thermodynamics of our planet. Solar radiation hits the earth with highest intensity near the equator. Because the earth's axis is tilted 23.5° with respect to the plane of its orbit, during part of the year the zone of maximum solar interception shifts northwards, towards the Tropic of Cancer, and during part of the year it moves southwards, towards the Tropic of Capricorn. Thus, the warm tropics form a belt around the equator from latitude 23° north to latitude 23° south, where the tropical heat generates rising, unstable air. As it climbs, the air condenses the moisture evaporated from the warm tropical seas and forests, and produces the heavy downpours that characterize the wet tropics. As it moves away from the equator at high altitudes, the air cools again and eventually starts descending towards the mid-latitudes, some 3000 km away from the equator both north and south. The air masses heat in their descent and, having lost their moisture during their tropical ascent, they become extremely dry. Thus, by contrast with the equatorial forests, the mid-latitude arid fringes that run alongside the tropical belt have a more stable atmosphere. These are the "horse" latitudes, where calm, dry air often dominates. Additionally, because of the stable atmosphere, not only are winds slack, but rainstorms seldom develop. For this reason most of the world's large deserts occur along the belt that separates the tropics from the temperate regions (Goudie and Wilkinson 1977).

CONTINENTALITY AND INLAND DESERTS

The sheer size of continents may be in some cases a direct source of aridity. Because most of the water in the atmosphere is ultimately derived from evaporation from the seas, there is often an aridity

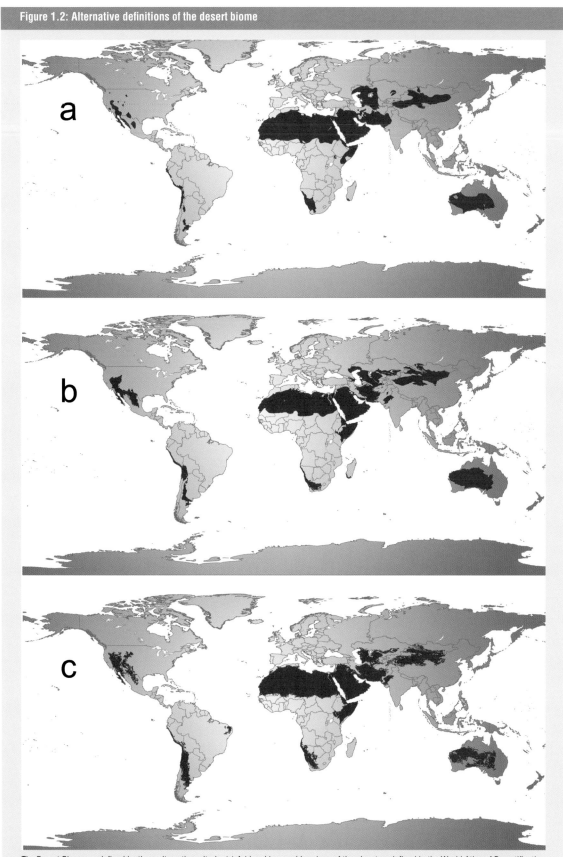

The Desert Biome, as defined by three alternative criteria: (a) Arid and hyperarid regions of the planet as defined in the World Atlas of Desertification (UNEP 1997); (b) global ecoregions of the world (Olson and others 2001; Appendices 1 and 2) that harbour desert vegetation; and (c) contiguous regions of extremely low vegetation cover derived from AVHRR satellite images of the world (USGS 2005).

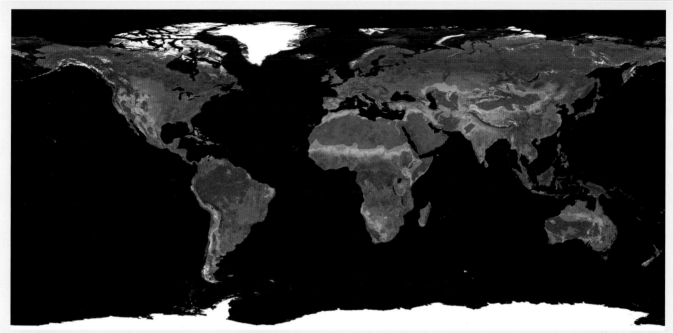

One of the first actions undertaken in the preparation of this report was to define the boundaries of the world's deserts. Deserts are commonly defined climatologically as the arid and hyper-arid areas of globe; biologically, as the ecoregions that contain plants and animals with clear adaptations for survival in arid environments; and physiologically, as large contiguous areas with low vegetation cover and ample extensions of bare soil. Overlaying the areas defined by each of the three criteria yields a composite definition of global deserts, occupying almost one-quarter of the earth's land surface — some 33.7 million square kilometres — inhabited by over 500 million people. The intensity of the red colour on the map indicates congruence in the three criteria: areas in intense red correspond to regions where the three criteria coincide, areas in intermediate red highlight regions where two criteria coincide, and areas in pale red show regions where only one criterion operates.

This analysis has also revealed that the population density in the desert cores (areas covered by all three definitions) is still low, while the edges of deserts (areas covered by only one or two definitions) are faced with higher pressures from human activities. Biological data also demonstrate that deserts are not simply barren lands but complex arrangements of diverse and fragile assemblages of species of flora and fauna. Located in the transition between deserts and semi-arid ecosystems, some of the desert fringes include several of the most endangered terrestrial ecoregions of the world.

	Figures in this report	Land area (1 000 km²)	Protected area[3] (% of area)	Hotspots[4] (% of area)	Human Population Total (1 000)	Human Population Density (persons/km²)	Distribution of land area (%) by degree of population pressure[5] Low	Medium	High
Deserts defined by:									
Aridity	Fig. 1.2a	25 714	4.5	9.0	354 976	14	94.1	3.82	2.1
Ecoregions	Fig. 1.2b	25 270	5.1	9.8	254 860	10	95.1	3.21	1.7
Landcovers	Fig. 1.2c	28 819	5.8	10.7	283 400	10	94.6	3.96	1.5
Deserts (strict) [1]	**Fig. 1.3**	**19 467**	**4.6**	**6.7**	**143 670**	**7**	**96.4**	**2.43**	**1.1**
Deserts (broad) [2]	**Fig. 1.3**	**33 688**	**5.5**	**12.1**	**502 232**	**15**	**92.9**	**4.77**	**2.3**
World Total		130 483	9.9	12.0	6 081 528	47	79.6	12.4	8.0

Note:
1. Deserts (strict) — The area classified as desert by all three definitions simultaneously.
2. Deserts (broad) — The area classified as desert by at least one of the three definitions.
3. Protected Area — The proportion of the area under environmental protection, following IUCN's definition, namely, "an area of land and/or sea especially dedicated to the protection and maintenance of biological diversity, and of natural and associated cultural resources, and managed through legal or other effective means." (Mulongoy and Chape 2004)
4. Hotspots — The proportion of the area occupied by biological "hotspots", defined as the earth's biologically-richest and most endangered terrestrial ecoregions (Mittermeier and others 1999)
5. Population pressure — Low population pressure: < 25 persons square kilometre; Medium population pressure: 25–100 persons per square kilometre; High population pressure: > 100 persons per square kilometre.

Source: San Diego Natural History Museum for image production, UNEP/GRID-Sioux Falls for table calculations

gradient in large continents: the land closer to the sea often receives a larger share of this ocean-derived water and, as air moves inland, it gets depleted of moisture and precipitation drops. Thus, regions lying deep within a continent may become deserts simply because air currents reaching them have already traversed vast land distances and lost most of the moisture they originally carried. Continentality is a major factor driving arid conditions in the Monte Desert in South America (see Figure 1.4), in the central deserts of Australia, in the Great Basin in North America, and especially in the central East Asian deserts, the Taklimakan and the Gobi.

COASTAL DESERTS: THE EFFECT OF MARINE UPWELLINGS ON DESERT DISTRIBUTION

The latitudinal explanation, however, is only partial. Around the mid-latitudinal belt, only the western side of continents is normally occupied by deserts, while the eastern side is covered by forests. The reason for this has to do with the global circulation of ocean currents: gravitation from the sun and the moon pulls air and water on the earth's surface

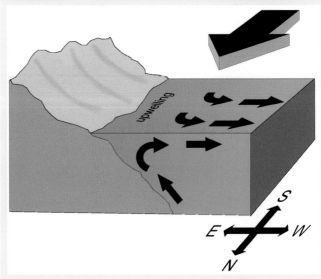
and tends to make them lag behind, relative to the earth's rotational movement. The gravitational drag is greatest in the equator, where the centrifugal speed of the earth is fastest. Thus, as the earth turns, ocean currents and winds flow in the equator from east to west, tugged by universal gravitation, forming the equatorial currents and the easterly trade winds. As the westbound surface waters move away from the continents, they pull cold, nutrient-rich waters to the surface that generate a cool, stable coastal atmosphere, with little evaporation from the sea and very low rainfall other than morning fogs (Figure 1.5). In the coasts neighbouring these oceanic upwellings, typical coastal fog deserts tend to develop, forming some of the driest ecosystems on earth. Thus, the large-scale circulation of the ocean is the main reason why coastal deserts are always found on the west side of continents, such as the Namib in Africa (Figure 1.6), Atacama in Chile, the Atlantic Coastal Desert of Morocco, or the deserts of Baja California (Figure 1.7).

RAIN SHADOWS AND TROPICAL DESERTS

Topographic heterogeneity also contributes to the formation of deserts, especially of those that occur outside the mid-latitude belts. In the tropics, for

Figure 1.4: Continentality and aridity

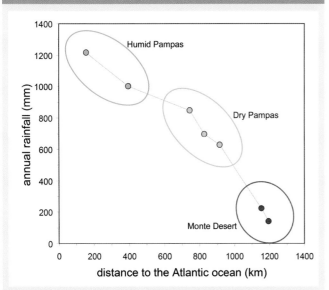

Following a transect along the flat Argentine Pampas from the Atlantic coast to the foothills of the Andes, mean annual rainfall decreases as a function of the distance to the coast at a mean rate of -1 mm/km. Because the high Andean Cordillera stops any moisture coming from the Pacific, the continental distance to the Atlantic coast is the single best predictor of precipitation. The colours indicate local weather stations in the humid Pampas (green), the dry Pampas (yellow), and, finally, the Monte Desert (red) in the extreme west.

Source: Servicio Meteorológico Nacional, Argentina; online statistics at http://www.meteofa.mil.ar/ for the following stations from east to west: Buenos Aires, Pergamino, Río Cuarto, Villa Mercedes, San Luis, Mendoza, and Uspallata

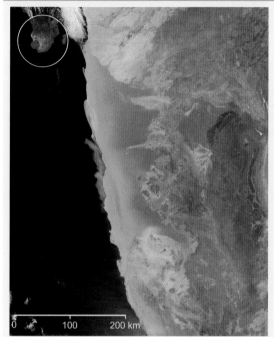

Figure 1.6: The Namibian coast

The consequences of the upwelling of cold, nutrient-rich water from the Benguela current near the coast of Namibia are clearly visible in the greenish-white plume of sulphur emissions produced by decomposing phytoplankton accumulated after the winter productivity peak. To the right, the dunes of the Namib desert bear witness to the dramatic aridity of this coastal desert.

Source: NASA 2004

example, when the moisture-laden tropical trade winds reach continental mountain ranges they cool as they ascend, condensing fog and drizzle that feed montane cloud forests. Once the winds pass the mountain divide, they start compressing and warming-up again in their descent, but, having left behind their original moisture, they become hot and dry. Thus, while the windward slopes of most tropical mountain ranges are covered by cloud forests, the leeward part, known as the "rain shadow" of the mountains, is covered by arid scrub. The rain shadow effect is largely responsible for many tropical arid lands that seem to defy the rule that deserts are only found in the earth's mid-latitudinal reaches, such as the Sechura Desert in Peru and Ecuador, the Caatinga scrub in equatorial Brazil, or the Tehuacán Valley desert in southern Mexico, a hotspot for cactus biodiversity. They are also responsible for some high-latitude cold deserts, such as the Great Basin, Patagonia, and the deserts of Central Asia.

DESERT LANDFORMS

The landforms of deserts, like those of high mountains and the polar areas, are much more visible than those of more vegetated landscapes. Bareness also allows much more active surface processes in all these areas, but in different combinations. Deserts suffer much more wind erosion than any other environment. Additionally, if slopes are steep and when the rain does fall, they also experience very fast water erosion. Desert landscapes come in two categories: (1) "shield" deserts and (2) "mountain-and-basin" deserts (Cooke and others 1992, Mabbutt 1977).

Shield deserts have developed on very ancient crystalline basements; that is, rocks that have been folded and faulted and hardened by heat and pressure over many millions of years. Granites, injected originally deep within the earth, have been unearthed by erosion and form steep-sided hills in many places (as at Uluru in Australia). The Sahara, the Arabian deserts, the southern African deserts, and the Australian deserts are in this group. Though very tough, the basement has been folded into gently-sloping swells and basins, and the basins have been filled over millions of years with sediments eroded from the swells, although these sediments have remained virtually unfolded themselves. They contain the best supplies of groundwater in the deserts, as in the northeastern and southern Sahara, and in Australia; and in some areas they are also rich in oil. Here and there recent volcanic rocks have overflowed at the surface, as in the Ahaggar and Tibesti mountains of the central Sahara. In their long lives these landscapes have experienced many different climates (partly because they were moved round the earth by continental drift) and many features formed in the different climates survive. There are even ancient glacial features in parts of Arabia and the Sahara; there are many more ancient river gorges, and ancient soils like silcretes or even laterites — ancient soils that formed under wet tropical conditions. The deep rotting of the rock in wetter times penetrated further in softer than harder rocks, and when the loose rotted material was stripped off, the uneven surface of the sub-soil landscape was revealed: a process appropriately called etching.

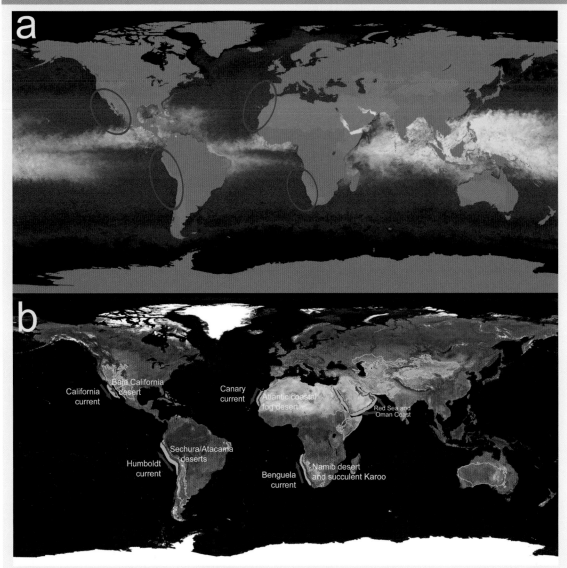

Figure 1.7: Sea surface temperatures and coastal deserts

a

b

California current
Baja California desert
Canary current
Atlantic coastal fog desert
Red Sea and Oman Coast
Sechura/Atacama deserts
Humboldt current
Benguela current
Namib desert and succulent Karoo

(a) Sea surface temperatures during 2–9 June 2001, measured by NASA's Moderate Resolution Imaging Spectroradiometer (MODIS). Cold waters are black and dark green; blue, purple, red, yellow, and white represent progressively warmer water. Note the plumes of cold water along the Pacific coasts of North and South America (Baja California and Peru), as well as along the Atlantic coasts of Africa in Morocco and Namibia. (b) Distribution of the coastal fog-deserts of the world and the associated cold ocean currents: the California, Canary, Humboldt, and Benguela streams. Other coastal and fog deserts are also found around the Arabian Peninsula, in both the Red Sea and the Oman coast, associated with the strong tidal currents in these narrow water bodies and straits.

Water is the main agent of erosion only on the few hills of the shield deserts, and cuts deep gullies on their edges. Elsewhere low gradients mean that water erosion is not very effective, and this leaves the field free to the wind: the great plumes of dust travel from the Sahara over towards Europe, southwest Asia and the Americas, removing much more sediment than do rivers from the same area, and have taken even more dust in recent geological periods. With the dust winnowed out, sand is left behind, and most of it collects in dunes, which cover 20–30 per cent of these landscapes. Some of the larger "sand seas" cover more than 300 000 square kilometres; their median size is 123 000 km^2.

Mountain-and-basin deserts are those in the much more recently folded and faulted rocks of the earth's active tectonic belts. Up-faulted mountains alternate irregularly with down-faulted basins. The American deserts, both North and South, are all of this kind, as are the deserts of Central Asia (where some of the basins, however, cover many hundreds of thousands of square kilometres). Water erosion in the mountains cuts deep, steep-sided valleys and gorges, and takes the debris out into the basins, where the broadening and shallowing of the ephemeral washes (called *arroyos* in the Americas and *wadis* in Northern Africa and West Asia) first cause the coarser debris to be dropped in broad "alluvial fans." Occasional extreme storms may carry huge boulders onto these as well. Further down, the alluvial fans coalesce into a long slope of finer alluvium — the *bajada*. Sand may be winnowed out of the alluvial deposits and form dunes (and in Central Asia, even a few sand seas). Only the finest debris (silt and clay) reaches the bottom of the basin, where it is deposited in ephemeral lakes or *playas*. The salts carried in the waters also accumulate there.

Climatic Variability and Rainfall Pulses

In deserts, rainfall events trigger short periods of high resource abundance which, despite the overall scarcity of rain, can saturate the resource demand of many biological processes for a short time. Thus, although deserts are often characterised by their mean climatologic conditions (as in the case of the Aridity Index described in the previous section) they are really driven by a succession of short pulses of abundant water availability against a background of long periods of drought. And, because rain storms are also frequently very localized, deserts are extremely patchy environments in their resource availability, both in space and in time. Rainfall pulses are really the driving force structuring desert ecosystems, and plants and animals have developed very specific adaptations to cope with ephemeral abundance, especially with regard to growth, population dynamics, and the cycling of organic matter and nutrients (Sher and others 2004). Within a desert, rainfall events may vary significantly from one pulse to the next: some spells may occur in winter, others in summer; some events may bring very little precipitation, others may bring intense showers; and the period between pulses may also vary substantially.

Each organism's response threshold is often determined by its ability to make use of moisture pulses of different durations and infiltration depths. For example, brief and shallow pulses have an important effect over surface-dwelling organisms with fast response times and high tolerance for low resource levels, such as soil micro-organisms. Short precipitation pulses are important to the survival of annual plants, but deep-rooted perennial plants may respond only to longer, more intense precipitation events. Thus, the diversity of pulses also promotes a diversity of responses in life-forms, migrations, or population cycles in different species. To a large extent, it is the heterogeneity of pulses that drives the surprisingly high biodiversity of desert ecosystems (Chesson and others 2004).

But what commands pulses in deserts? Why are seasons, and even decades, so different from each other, and often so unpredictable? To a large extent, pulse-type variations in desert environments are linked to global atmospheric and oceanic phenomena. Large-scale drivers of regional precipitation patterns include the position of the jet streams, the movement of polar-front boundaries, the intensity of the summer monsoon, El Niño Southern Oscillation events, and even longer-term ocean cycles, such as the Pacific Decadal Oscillation (Loik and others 2004). Driven by these large-scale forces, the intensity of mid-latitude continentality, ocean upwellings, and rain shadows — the major factors modulating the distribution of arid lands — is not constant but may vary from one year to the next. As a result, the intensity and frequency of rain pulses at a local scale may vary substantially with time, and in a seemingly unpredictable fashion.

The influence of large-scale drivers on local desert conditions was noted many years ago by the fishermen and the farmers of the coastal desert of Peru, who realized that during some years the normally cold waters of the Pacific became warmer. In these years, they noted, the abundance of sardines decreased but abundant rainfall soaked the land and made the desert flourish. Because this phenomenon was normally observed around the month of December (a time of the year in which Christians commemorate the birth of the Christ child — *El Niño* in Spanish), they called the phenomenon

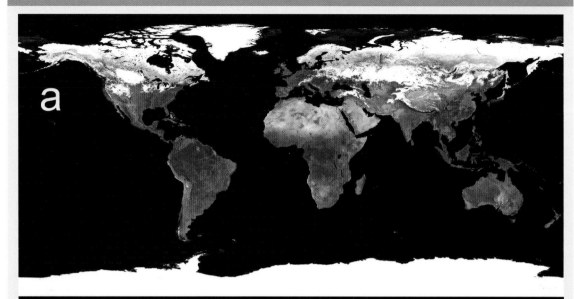

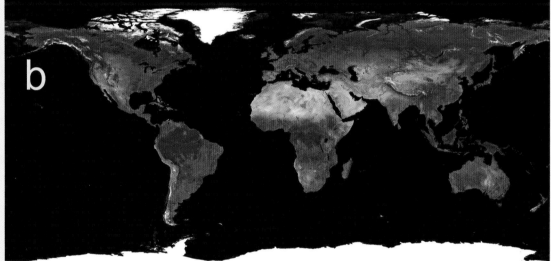

While snow and ice cover a large proportion of the temperate terrestrial areas during the northern hemisphere winter (January 2004, top) and all but disappear in summer (July 2004, bottom), the advance and retreat of the ice sheet in the southern hemisphere is much less pronounced because of the buffering effect of large oceanic masses. A similar effect occurred during the Pleistocene glaciations, when the ice sheet covered most of the northern hemisphere, forcing deserts north of the equator to retreat into the tropics. In the southern hemisphere, in contrast, because of the lower landmass and higher oceanic influence, the advance and retreat of the glaciers was less pronounced.

Source: NASA 2004

During the warm early to mid-Holocene (8 000–5 000 yBP), the global climate that resulted from glacial retreat brought an increase in the intensity of the monsoon throughout the sub-tropical arid lands. Lake Chad became a freshwater inland lake bigger than today's Caspian Sea, in an area that has again become a complete desert. Tropical forests and dry woodlands around the equator expanded north and south, while deserts moved into the mid-latitudes. During that period, the southern Sahara and the Sahel were much wetter than today, with extensive vegetation cover, thriving animal communities, and numerous human settlements.

Sometime between 6 000 and 5 000 yBP, there was again a transition to more arid conditions. Mesic vegetation communities disappeared rapidly, lake levels declined dramatically, and highly mobile pastoralist cultures started to dominate and replace sedentary lacustrine and riparian traditions. The Liwa region of the United Arab Emirates, for example, experienced phases of sand deposition that lead to the formation of a large (up to 160 m high) mega-dune. A similar transition towards more arid conditions occurred in North America, where the Holocene brought the arrival of Mojave, Chihuahuan and Sonoran desert scrub elements from the south, such as the agaves, cacti, ocotillos (*Fouquieria*), and creosote bushes that characterize the area today.

An explanation for these climatic variations is that changes in incoming solar radiation, associated with slow shifts in the Earth's orbit, enhanced the strength of the summer monsoon rains at the beginning of the Holocene. These rains, in turn, increased the extent of vegetation cover and wetlands, and this had two major effects — a reduction in surface albedo (reflectance) and an increased ability to recycle water back to the atmosphere through evapotranspiration. Both effects helped fuel the monsoons with additional energy and moisture, increasing the summer rains. In Africa, the climate-vegetation system maintained a "green Sahara" climatic regime through the middle Holocene, when a sudden transition occurred to a "desert Sahara," the regime that we know at present. The aridization trend of the mid-Holocene fed back into the deserts themselves by decreasing vegetation cover, reducing local inputs of moisture into the atmosphere, and further increasing the dry conditions.

MOUNTAIN SKY-ISLANDS AND CLIMATES OF THE PAST

When the ice sheets started to retreat, some 20 000 years ago, most of the temperate flora and fauna slowly migrated back into higher latitudes and the Desert Biome gradually expanded across the mid-latitudes to its current extent. A subset of the temperate biota, however, stayed behind in the rugged and cool mountain ranges that emerge from the desert plains. Establishing higher-up with each passing generation as the climate warmed, the ice-age organisms were able to persist in the cool mountain environments, where conditions are similar to the ones they had enjoyed in the lower plains during the Ice Ages. As they ascended into the isolated desert mountains, the communities of the desert "sky-islands" became separated from other mountains by harsh desert plains. Like prehistoric castaways,

the Ice Age species now survive high-up in the cool refuges of the desert mountains; a biological memory of past evolutionary history subsisting high-up in the mountains like ghosts of climates past. And, because they have been reproducing in isolation for 15 000–20 000 years, many of their populations have developed unique genetic traits and have evolved into new species. Thus, in a similar fashion to evolution in remote oceanic islands, the biota of the desert sky-islands is composed by a large number of endemic species and has immense value for biological conservation (Axelrod 1950).

As the effect of the Ice Ages was more severe in the northern hemisphere, which is mostly covered by continental land masses, than in the more oceanic southern half of the globe (Figure 1.9), most of these Pleistocene montane relicts are found north of the equator. In North America, where the desert relief is highly folded, mountainous sky-islands dapple the central part of the Sonoran and the Chihuahuan Deserts, and of the Great Basin. All these ranges contain endemic pines, oaks, madrones, and chaparral species, remnants of the "Madro-Tertiary" flora, a unique temperate ecosystem that covered much of the now-dry North American deserts during the last six million years.

In Africa, similar relict mountains emerge from the harsh Saharan plains: near the Mediterranean coast, the Atlas Mountains in northern Morocco shelter rich pine and oak forests. Further south, the Ahaggar and Tassili-n-Ajjer ranges of south-eastern Algeria and the Aïr massif in northern Niger harbour a number of endemic and rare Mediterranean species such as the tarout, the wild olive, and the Saharan myrtle. To the east, the Tibesti mountains in southern Libya hold some Mediterranean as well as some tropical relicts. These Saharan mountains also provide prime habitat for migratory birds and a key refuge for threatened wildlife.

In the Somali Peninsula, and across the Gulf of Aden in the southern part of the Arabian Peninsula, high mountain ranges shelter similar temperate relicts: along the northern Somalian coast, in the tip of the Horn of Africa, the Somali Montane Woodlands thrive along the coastal ranges fed by

Woody desert trees, such as acacias, cannot store much water in their trunks but many of them evade drought by shedding their leaves as the dry season sets in, entering into a sort of drought-induced latency. Many of these desert species also have deep taproots that explore deep underground water layers. Other trees have convergently evolved a mixture of these strategies: they can store water in gigantic trunks and have a smooth bark that can do some cactus-like photosynthesis during dry periods; but, when it rains they produce abundant green leaves and shift their metabolism towards that of normal-leaved plants. This group is formed by trees with famously "bizarre" trunks, such as the African baobab (*Adansonia*), the Baja-Californian Boojum-tree (*Fouquieria columnaris*) and elephant-trees (*Bursera* and *Pachycormus*), and the South African commiphoras (*Commiphora*), bottle-trees (*Pachypodium*), kokerbooms (*Aloe dichotoma*, Figure 1.12) and botterbooms (*Tylecodon*).

A third group of plants, the "true xerophytes" or true desert plants, have simply adapted their morphology and their metabolism to survive extremely long droughts. These species have remarkably low osmotic potentials in their tissue, which means that they can still extract moisture from the soil when most other plants cannot do so. True xerophytes, such as the creosote bush (*Larrea*), are mostly shrubs with small, leathery leaves that are protected from excessive evaporation by a dense cover of hairs or a

thick varnish of epidermal resin. Their adaptive advantage lies in their capacity to extract a fraction of soil water that is not available to other life-forms. However, because their leaves are so small and protected from transpiration, their gas-exchange metabolism is very inefficient during rain pulses when moisture is abundant. In consequence, these species are extremely slow growers, but extremely efficient water users and very hardy.

Finally, one of the most effective drought-survival adaptations for many species is the evolution of an ephemeral life-cycle. A short life and the capacity to leave behind resistant forms of propagation is perhaps one of the most important evolutionary responses in most deserts, found not only in plants but also in many invertebrates. Desert ephemerals are amazingly rapid growers capable of reproducing at a remarkably high rate during good seasons, leaving behind myriad resistance forms that persist during adverse periods. Their population numbers simply track environmental bonanzas; their way to evade critical periods is to die-off, leaving behind immense numbers of propagules (seeds or bulbs in the case of plants, eggs in the case of insects) that will restart the life cycle when conditions improve. These opportunist species play an immensely important role in the ecological web of deserts: myriad organisms, like ants, rodents, and birds, survive the dry spells by harvesting and consuming the seeds left behind by the short-lived ephemeral plants. Granivory (the consumption of seeds) and not herbivory (the consumption of leaves) is at the base of the food chain in most deserts, as those few plants that maintain leaves during dry spells usually endow them with toxic compounds or protect them with spines. The onset of rainy periods brings to the desert a reproduction frenzy of desert ephemerals, and a subsequent seed-pulse that drives the entire food web for years.

From the information above it can be seen the survival strategies of desert plants are classifiable along a gradient ranging between two extreme categories: (a) adaptation for quick use of ephemerally abundant resources, or (b) adaptation for the efficient use of poor but more permanent resources (Shmida 1985). The first category, typically exhibited by desert ephemerals, represents

Figure 1.12: Fleshy-stemmed trees

The fleshy stems of the kokerbooms (*Aloe dichotoma*)— one of Africa's many giant, fleshy-stemmed trees — are true landmarks in the otherwise barren landscape of the Namib Desert.

Source: Patricia Rojo

a "maximum variance" behaviour that consists essentially in tracking environmental variation, while the second category, exhibited by true xerophytes and cacti, is a "minimum variance" behaviour that consists in adapting to the worst possible conditions. Drought-deciduous perennials and grasses represent a compromise between these two extreme behaviours. Attributes necessary for the quick use of water include rapid growth (often at the cost of low water-use efficiency) and abundant seed production. Attributes for survival with little water include high water-use efficiency, slow growth, and passive cooling. Drought deciduousness, as an intermediate strategy, requires the capacity to shed leaves and to quickly recover them when moisture conditions improve.

The survival strategies of desert plants present some of the most striking cases in nature of evolutionary convergence: plants from widely different families and from divergent evolutionary origins have developed, in the different deserts, life-forms so similar that it is sometimes difficult to tell them apart. Such is the case of the succulent cactoid growth form, evolved in Africa from the families Euphorbiaceae, Asclepediaceae, and Aizoaceae, and in the Americas from the family Cactaceae. Similarly, bottle trees in Africa evolved from the families Apocynaceae, Aloeaceae, and Crassulaceae, while in the New World they belong to the Fouquieriaceae, Anacardiaceae, and Burseraceae.

ADAPTATIONS OF ANIMALS TO ARIDITY
Behavioural adaptations
To the physiological, anatomical, and morphological adaptations of plants, animals can add adaptive behaviour. Many birds and most large mammals, like pronghorn antelopes or wild sheep, can evade critical spells by migrating along the desert plains or up into the mountains. Smaller animals cannot migrate such long distances, but regulate their environment by seeking out cool or shady places. In addition to flying to other habitats during the dry season, birds can reduce heat loads by soaring. Many rodents, invertebrates, and snakes avoid heat by spending the day in caves and burrows, and procuring food during the night. Even diurnal animals may reduce their activities by resting in the shade during the hotter hours of the day. Fossoriality, a lifestyle based in burrows, is the

Figure 1.13: The sidewinder snake

The characteristic crawling patter of the sidewinder (*Crotalus cerastes*) leaves a tell-tale trail in the dunes of the Gran Desierto. Its perfect matching to the colour of the sand protects it from predators such as kestrels and falcons.

Source: Patricio Robles-Gil

norm for small animals in all deserts, as it allows them to stay away from the gruelling heat during the hotter part of the day and it also provides them a warm refuge during the cold desert nights. Additionally, humidity inside burrows (ca. 30–50 per cent) allows desert animals to preserve water. When the normal mechanisms to keep body temperature within acceptable limits fail, many small rodents and some desert tortoises (*Testudo*) salivate to wet the chin and throat and allow evaporative cooling. Such mechanisms have a high cost in water and are used only as emergency measures to prevent death.

At dawn, the dry desert ground may approach freezing temperatures and at midday it may heat up into an 80°C inferno. A few inches above the ground, variations in air temperature are much less pronounced, and, just a few inches below the surface, underground temperatures are almost constant between day and night. For this reason, thermoregulation is a particularly challenging problem for small surface-dwellers and especially for reptiles, which cannot regulate their body temperature metabolically. Most desert reptiles have developed peculiar ways of travelling over hot sandy surfaces. Side-winding, a form of lateral movement in which only a small part of the body is in contact with the surface, is employed by many sand snakes (Figure 1.13). Many lizards and some ground birds avoid overheating by running rapidly over the hot desert surface while maintaining their bodies well separated from the ground (Safriel 1990). Some lizards assume an erect, bipedal position when running, while others regulate their contact with the hot desert pavement by doing "push-ups" with their forelegs.

Many large mammals that cannot avoid being in the sun during a large part of the day orient their bodies so as to reduce the incidence of the sun's rays. By standing upright, ground squirrels reduce solar incidence upon their bodies. The African ground squirrel *Xerus inauris* even orients towards the sun and shades itself with its tail when foraging (Figure 1.14). The jackrabbit *Lepus californicus* warms its body in the early morning by exposing its large, highly vascularised ears perpendicular to the sun's rays, using them as a form of solar collector. Similarly, it cools at midday by keeping in the shade and putting the ears parallel to the incoming solar radiation, thus minimizing exposure while keeping the same radiative surface.

Figure 1.14: The African grand squirrel

Using its tail like a parasol, the African ground squirrel (*Xerus inauris*) protects itself from the sun in the Namib Desert.

Source: Patricio Robles-Gil

Figure 1.15: Ostriches in the Namib

Despite the gruelling heat and extreme drought, a family of ostriches (*Struthio camelus*) survives and thrives in the Namib Desert.

Source: Patricio Robles-Gil

Morphological and anatomical adaptations

In mammals, desert fur coats are short, hard and compact, but at the same time well-ventilated, to allow sweat to evaporate directly from the skin. Birds, in addition, can fluff or compact their feathers to regulate heat exchange. In the ostrich, a desert dweller, the uncovered head, throat, legs and abdomen allow for radiation and convection cooling, while the feathers on the back protect the larger part of the body from direct solar radiation (Figure 1.15). Bipedalism, a common trait in small desert mammals such as kangaroo rats, allows for fast travel in open spaces and also keeps the body separated from the extreme temperatures of the ground surface. Indeed, bipedal desert rodents use open microhabitats much more frequently than their quadrupedal relatives, who restrict their activities to sheltered habitats.

Sand-dwellers have evolved several traits that allow them to survive in dunes, including fleshy foot-pads in camels, scaly fingers in certain lizards and digital membranes in some geckoes. Additionally, camels have long dense lashes that protect their eyes and they can close their nostrils to protect them from wind-blown dust. Many snakes have upwardly-turned nostrils that allow them to burrow rapidly in loose sand; others are flat and can bury laterally. Many other reptiles also show adaptations that protect their eyes, nose, and ears from sand

and dust, and many insects have specially-adapted legs that allow them to bury themselves rapidly and to walk efficiently on hot sand.

Physiological adaptations

The most basic physiological problem of desert animals is to maintain their water balance by maximizing water intake and/or minimizing water loss. In deserts, free-standing water is scarce, found only in isolated oases and reservoirs. Camels and wild asses, for example, can drink large quantities of water in a very short time causing a dramatic dilution of the bloodstream, sufficient to cause death in other animals. In coastal deserts, animals obtain water by licking fog-drenched rocks. Desert amphibians can absorb water through the skin from humid underground dens by accumulating urea in their blood and raising its osmotic pressure.

Most herbivores, like eland and oryx, obtain water from the foliage of the shrubs that compose their diet, often feeding at night when the plants are turgent. Some succulent plants have high salt contents, toxic compounds, or spines that deter their consumption. Herbivores, however, have found their way around these obstacles: some reptiles and birds have developed efficient salt-excreting glands, and many mammals have kidneys that can cope with salty water. White-throated packrats (*Neotoma albigula*), which feed almost exclusively on juicy cacti, have metabolic adaptations to prevent poisoning from the oxalates contained in these plants.

Animals lose water through urine, faeces, respiration, and transpiration. Desert rodents have kidneys that are capable of producing highly concentrated urine, with an electrolyte concentration many times higher than that of blood plasma. Reptiles, birds and insects excrete uric acid, which requires less water, and sometimes complement the excretion process with specialized excretion from salt glands. Amphibians produce little urine, and can store large amounts of urea within their bodies, drastically reducing water loss. In droughts, some rodents can produce dry faeces, efficiently reabsorbing liquids in the rectum.

Metabolism produces CO_2 and water as by-products of respiration. In most animals, this metabolic water is exhaled through the lungs, but many desert animals, including invertebrates, reptiles, and mammals, possess physiological and anatomical adaptations to reduce respiratory water loss, including modifications in the morphology of the nasal passages and the capacity to reabsorb water along the respiratory tract. One of the most extreme examples of this is given by the kangaroo rat (*Dipodomys*), which can survive on a diet of perfectly dry seeds.

In addition to the mechanisms that reduce water loss, many desert animals are extremely tolerant to dehydration, a condition that causes a fatal increase in blood viscosity in non-desert dwellers. In order to achieve this, camels, for example, are able to loose water selectively from tissues other than blood. In contrast, desert amphibians are tolerant to increased fluid viscosity, and some reptiles can excrete excess electrolytes through urine and salt glands, avoiding the thickening of the blood as they dehydrate. A problem related to dehydration is that of temperature regulation. In smaller animals the high surface area-to-body volume ratio makes sweating a dangerous enterprise and panting is the most common method of cooling. Even larger animals that usually sweat, like the oryx, begin to pant when their body temperature exceeds 41°C.

Nocturnal hypothermia, exhibited by some large mammals like the eland, allows them to reduce their metabolic rate and to exhale air with less humidity during the night. Diurnal hyperthermia allows animals to reach body temperatures that would be normally lethal for non-desert vertebrates and to save on water needed to prevent overheating. Camels and elands, for example, can reach body temperatures of 44°C with no harmful consequences, and save as much as 5–10 litres of water during extremely hot days. Hyperthermal species have a special disposition of veins and arteries that allows their brains to remain at a temperature lower than that of the overheated body.

Like ephemeral plants, many smaller desert animals can also evade drought by entering into a dormant phase: desert butterflies and grasshoppers thrive in huge numbers when conditions are good and survive dry spells in the form of eggs or pupae. Spade-foot toads (*Scaphiopus*) spend most of their lifetime buried in dry mud and become active only after rains refill their ephemeral pools. Many other organisms go into some form of torpor during dry periods.

INTERACTIONS BETWEEN SPECIES

The harsh conditions of desert ecosystems has promoted the evolution of a complex set of relations among desert organisms, a surprising number of which are positive interactions (Cloudsley-Thompson 1996). Desert shrubs in general and woody legumes in particular, create microhabitats that are critical for the survival of other species. Small animals seek the shade of desert trees and shrubs, birds find refuge and nesting sites in their canopies, and many small plants recruit their juveniles under the nitrogen-rich canopy of desert woody legumes such as acacias, carobs, and mesquites. Because of their crassulacean acid metabolism, desert succulents such as agaves, aloes, and cacti

Figure 1.17: Seed dispersal

A Gambel's quail (*Callipepla gambelii*) forages on the ripe, red, juicy fruits of the saguaro. After digesting the sweet pulp it drops the seeds, often under the protective shade of nurse plants, where the cactus will be able to germinate and establish.

Source: Patricio Robles-Gil

are poor thermoregulators as young seedlings, and cannot survive the harsh ground-level midday temperatures. For this reason, they can successfully germinate and establish only under the protective shade of shrubby "nurse plants" that act as true cornerstone species in desert conservation (Figure 1.16). If the desert trees and shrubs are cut, all the accompanying biota soon disappears.

Additionally, many desert plants have very specific requirements in terms of their pollinators and seed dispersers (Figure 1.17). Although some desert ephemerals are truly unspecific in their requirements and produce thousands of seed with only wind-pollination, the slow-growing desert perennials are frequently highly specialized in their reproductive habits, and depend strictly on co-evolved animals to help them out in their sexual and reproductive processes. Many African cactoid plants (euphorbs and asclepias) produce foul-smelling flowers that attract carrion insects as pollinators. New World giant cacti and agaves produce sugar-rich nocturnal flowers that engage the pollinating services of nectar-feeding bats, while the sweet pulp of their fruits lures birds to disperse

Figure 1.16: Nurse plants

Giant saguaro cacti (*Carnegiea gigantea*) in the Pinacate mountains of the Sonoran Desert establish and grow under the protective shade of a palo verde (*Parkinsonia microphylla*), a legume species that generates true islands of fertility in the harsh desert environment.

Source: Patricio Robles-Gil

the seeds miles away. The red tubular flowers of many desert shrubs attract hummingbirds and giant sphinx-moths.

"Co-evolution" is a term evolutionary theorists use to describe ecological intimacy, when the evolution of one organism is shaped by and in turn shapes the evolution of another (Ehrlich and others 1988). It is interesting to consider the evolutionary trajectories of desert biota and of human beings in that light. Relative to the time-scales of the geologic and atmospheric processes that created desert conditions, the advent of humans is a very recent event. Nonetheless, the effect of humans on deserts — and of deserts on humans — is pronounced.

Desert landscapes and desert biota have had profound effects on human cultural evolution. Humans display remarkable behavioral and cultural adaptations to the aridity and unpredictability of deserts, and traditions derived there have influenced human and biological communities far beyond the desert edge; that the three "religions of the book" had their origins in these environments well-illustrates that fact (see Chapter 2). Plants and animals from these harsh landscapes have also played an important role in the evolution of modern human societies. Dryland biota provided much of the "raw material" for species that could and did become domesticated, which helped usher in the dawn of pastoral and agricultural societies. The early domestication of ungulates (cattle, sheep, and goats) began in the drylands of West Asia, on the edge of the Arabian Deserts, some 9 000 years ago (Davis 2005), and the domestication of llamas and alpacas took place in the Andean Puna of South America some 6 000 years ago just north of South America's "arid diagonal" formed by the Atacama, Dry Puna, and Monte deserts (Table 1.3). In many regions of the world, dryland annuals have been at the base of the plant domestication process and drylands have been the cradle of agricultural societies. The first records of cultivated wheat and barley (two dryland ephemerals) come from the Fertile Crescent of West Asia some 7-9 000 years ago. In the American Continent the first agricultural records come from the Tehuacán Valley in southern Mexico, a hot tropical dryland where corn and squash (two annual, drought-tolerant fast growers) were first domesticated some 6 000 years ago. Not too long after that, gatherers in the Andean Puna started domesticating two other dryland ephemerals: the quinoa (*Chenopodium*, a fast-growing annual) and the potato (*Solanum*, a tuber ephemeral).

Table 1.3: Domestic mammals evolved from wild dryland ancestors						
Domestic form		**Wild progenitor**		**First domestication**		**Distribution of wild progenitor**
Common	**Scientific**	**Common**	**Scientific**	**Date**	**Place**	
Perissodactyla						
Donkey	*Equus asinus*	African ass	*Equus africanus*	4000 BC	Egypt	North Africa, possibly West Asia
Artiodactyla						
Llama	*Lama lama*	Guanaco	*Lama guanicoe*	4000 BC	Peruvian Andes	South American drylands
Alpaca	*Lama pacos*	Guanaco	*Lama guanicoe*	4000 BC	Peruvian Andes	South American drylands
Dromedary	*Camelus dromedarius*	Dromedary	*Camelus* sp.	3000 BC	West Asia	Asia, possibly North Africa
Bactrian camel	*Camelus bactrianus*	Bactrian camel	*Camelus ferus*	3000 BC	Central Asia	Central Asia
Goat	*Capra hircus*	Wild goat	*Capra aegagrus*	7000–8000 BC	West Asia	West Asia
Sheep	*Ovis aries*	Mouflon	*Ovis orientalis*	7000–8000 BC	West Asia	West Asia

Source: FAO 2002

Old and young desert landscapes, landform/soil/hydrology interactions

Soil is the earth's living skin — the stuff from which plants grow, shelter for myriad animals and microorganisms, and the surface that partitions rainfall into run-off, "green" water held in the soil accessible to plants, and drainage to groundwater. Life, the atmosphere, surface and groundwater, soils and landforms have evolved together.

Soils are formed by climate, topography, and life acting on the parent material over time. They are not all the same; each individual soil shows a record of its development in the soil profile — seen in a vertical slice down from the surface. Because rain is rare in deserts, chemical weathering of the parent material and leaching of weathering products such as clay, lime and soluble salts hardly takes place; rather, evaporation drives the upward movement of water and dissolved salts which accumulate at, or close to, the surface.

Desert terrain determines the distribution of soil and water: in steep lands, erratic rains and erosion restrict the soil to a thin, patchy cover; light showers merely wet the surface and the water evaporates where it falls; rare torrents produce flash floods that carry water, mud and dissolved salts to footslopes and depressions.

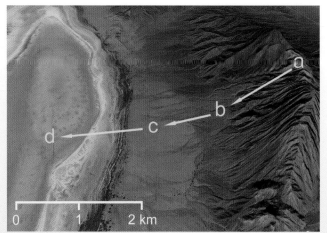

The desert topographic sequence clearly visible at the Laguna Salada in the Mexican Sonoran Desert. The upland mountains (a) discharge their run-off onto alluvial fans of the terraced foothills (b); downstream, the water flows in the bajada (c) coalescing into larger streams dotted by mesquite and other phreatophytes (plants that extract water from the deep aquifer), and finally the run-off and drainage water with all its accumulated salts arrives in the valley bottom playa (d), where it evaporates leaving the salts behind. Thus, the playa salt flat is fringed by a belt of halophytes, or salt-tolerating plants. Along these desert topographic sequences, there is an inverse texture effect as compared to wetter regions: soil texture becomes finer towards the valley bottom where the clay is transported by water. The hillslopes are rocky, the foothills are stony, the bajadas are formed by sand and loam, and the playas by clay.

Source: Google Earth image browser

Where soil and water accumulate, vegetation takes hold — generating organic matter that, in turn, sustains animals and microorganisms. Decomposition of organic matter produces CO_2 and organic acids that dissolve carbonate and silicate minerals. Materials in solution — soluble salts, gypsum, lime, silica, manganese, iron — may then be redistributed within the soil profile and in the landscape.

Time, as a soil-forming factor, denotes no absolute age but refers to the accumulated time in which soil formation could take place. Desert soils may be very old in years, but they usually show only weak signs of alteration of the soil parent material, because chemical weathering and biological activity are only spasmodic. Evidence of redistribution of sparingly soluble materials is generally more pronounced in the deserts of old continental shield areas than in younger formations. Thick, hard crusts of silcrete, lime and/or gypsum characteristic of old desert areas were once formed at some depth below the surface, but have been exposed by erosion; areas into which deserts have expanded in more recent times often preserve these relics of past wetter conditions.

Desert soils
The most extensive soils of deserts are the stony, rocky soils of the uplands, windblown sands, and other soils that have only surficial profile development. Although typical of true deserts, they are not strictly confined to deserts and may be found also in other drylands. These soils include *Leptosols* (Greek *leptos*, thin), which are shallow soils over hard rock, *Arenosols* (Latin *arena*, sand), developed in loose sand, and *Regosols* (Greek *reghos*, blanket), developed in other unlithified materials. They exhibit only a surface layer which, in deserts, is hardly different from the underlying parent material.

The most characteristic desert soils are those that exhibit significant accumulations of soluble salts: gypsum, calcium carbonate, or silica. These include *Solonchaks* (Russian *sol*, salt; *chak*, salty area), characterized by high levels of soluble salts which accumulate naturally in closed depressions; Solonetz (Russian *sol*, salt; *etz*, strongly expressed), highly alkaline, sodium rich soils which are very slippery when wet; *Gypsisols*, or gypsum soils; *Calcisols* (Latin *calcarius*, chalky or limy), characterized by accumulation of calcium carbonate, and *Durisols* (Latin *durus*, hard), showing a hard pan cemented by silica.

Accumulation of salts in desert soils affects plant growth in two ways: (1) the salts aggravate drought stress because electrolytes create an *osmotic potential* that opposes uptake of water by plants; (2) dissolved sodium depresses the uptake of potassium and calcium, and magnesium depresses potassium. Thus, the vegetation of *solonchacks* is highly specialized: several species — called halophytes or salt-plants — have evolved fleshy leaves with large vacuoles in their cells, where they dispose of the excess ions that are taken up by the roots, and characteristic red algae develop in the hypersaline substrates of the valley-bottom *playas*.

Dissolution of calcite ($CaCO_3$) and its subsequent precipitation in a deeper calcic (soft) or petrocalcic (hard) accumulation layer is controlled by (a) the CO_2-pressure of the soil air and (b) the concentrations of dissolved ions in the soil water. The mechanism is straightforward: the partial pressure of CO_2 is high where root activity and respiration by soil microorganisms produce CO_2, so calcite in the topsoil dissolves as Ca^{2+} and HCO_3^- ions which move down in percolating water. Evaporation of water and a decrease in CO_2-pressure deeper in the soil (fewer roots and less soil organisms) cause saturation of the soil solution and precipitation of calcite. Calcite, like gypsum, does not precipitate evenly within the soil matrix; fissures, root channels and burrows that are connected with the outside air act as ventilation shafts in which the CO_2-pressure is much less than in the soil matrix. When carbonate-rich soil water reaches such a channel, it loses CO_2 and calcite precipitates on the channel walls. Narrow channels become entirely filled with calcite *pseudomycelium*. Other characteristic forms of calcite accumulation are soft or hard *nodules, pendants* or *beards* below stones, and platy layers of hard calcrete.

Where is the water?

Landforms and soils, jointly, determine hydrology and, in turn, ecology.
Variations of soil and terrain from steep and rocky land to mobile dunes to fans and playas, with soils ranging from shallow patches of earth on hard rock to thick sand or clay, many with hard cemented layers close to the surface, some with a high concentration of salt at the surface — all determine the distribution and availability of water and footholds for life in deserts. The variety of life-forms and life strategies has evolved to exploit these specialist opportunities.

Rainfall in deserts can be heavy; it just happens infrequently. This means that plant cover is usually sparse so the rain batters an unprotected soil surface. Most of the soils yield rapid run-off — thin, rocky soils of the steeplands and thicker soils of footslopes slake on wetting and do not allow rapid infiltration. Run-off causes erosion; it quickly gathers as muddy torrents that carve steep *wadis* or *arroyos* in the uplands and, then, deposit fans of alluvium where they emerge on the footslopes. In tectonically active regions, these landforms are commonly disrupted; sections of fans and floodplain may be raised as terraces along the flanks of mountain chains, valleys are offset along fault lines. These features are very visible, because there is little vegetation to conceal them, and they remain fresh because erosion and deposition by water is infrequent.

Streams are short-lived. On gentler slopes, the water soaks into its own alluvium, maintaining a groundwater table that may appear at the surface in the lowest parts of the landscape as springs, oases and lakes. Where the water-table is close to the surface, evaporation drives the upward movement of water and, with it, dissolved salts that accumulate as a surface efflorescence, salt lakes and saline groundwater. Towards the centre of any landlocked basin, surface and groundwaters tend to become increasingly saline though they may be fed by rain and snowmelt on the surrounding mountains.

The patterns of soils and vegetation reflect this hydrological and salinity gradient:

Uplands: Shallow, patchy soils (Lithosols); may be desert or not desert, even fed by snowmelt; generate run-off and deep drainage to groundwater — both fresh. In desert, vegetation is very sparse because water does not remain long enough to be useful.

Terraces: Terraces, often formed by the accumulation of coarse rocks, are found at the foothills of the desert mountains. Older terraces carry more mature soils, maybe with subsoil layers of carbonate or gypsum accumulation (Calcisols or Gypsisols) that may impede water percolation; subject to erosional run-off. Where there is reasonably regular rainfall, the thicker topsoil may support steppe or thorn/cactus scrub.

Bajadas (coalescing alluvial fans that may bury lower terraces in fresh alluvium): Variation in soil texture depending on the force of water flow, finer materials deposited on gentler slopes furthest out across the fan; surface water soaks into the alluvium. Vegetation follows the more regular water courses and deep-rooted perennials may exploit shallow groundwater.

Marl plain or *playas*: Lacustrine flats beyond the toe of the fans; springs where the water-table intersects the surface, commonly upwelling of deep seepage from the surrounding uplands providing a rich habitat; evaporation concentrates the seepage leading to a salinity gradient extending across the plain; least-soluble salts crystallize first so there is a sequence of carbonates (Calcisols), sulphates (Gypsisols) and, finally, the very soluble chlorides (*Solonchak/Solonetz*); vegetation has to be increasingly specialized.

Box authors: David Dent and Paul Driessen

That biota of drylands would be a source of such innovation is not surprising, given the life history of those plants and animals. Arid-land herbivores, in particular desert ungulates, are extremely hardy. They can use water very efficiently, they can withstand long periods without drinking, and when forage is plentiful they can quickly convert plant material into animal protein with very high efficiency. Furthermore, many of them are migratory and move naturally in herds following a leader, looking for new foraging grounds, and socially protecting themselves from predators. For all these evolutionary reasons, ungulates native to drylands were ideal candidates for domestication: hardy animals, efficient foragers, and amenable to shepherding, as social aggregation is a natural behaviour for them. Some of the same factors that made wild goats, mountain sheep, or guanacos evolutionarily adapted to desert environments are what drove early hunter-gatherers to start breeding their offspring and selecting them for desirable domestic attributes. As with desert ungulates, the same traits that have made some desert annuals apt to survive and thrive on ephemeral water pulses are what make them so apt for agriculture: fast growth, short life cycle, and the capacity to direct most of their metabolic budget towards the abundant production of seeds. Because dryland ephemerals grow so fast and produce so much seed in just a few weeks, they grow at an amazingly fast rate when planted at the desert's edge and make ideal grain plants, especially cereals and pulses.

The effect of humans on the ecology and the evolutionary trajectory of deserts can be similarly pronounced. The following ecological "anachronism" provides an illustrative example: some desert plants have seed dispersal mechanisms that reflect the existence of seed dispersers that are no longer present. Trees like the mesquites (*Prosopis*), for example, have pods with nutritious sweep pulp and extremely tough seeds which need intense scarification in order to germinate. Similarly, the tough seeds of the prickly pears (*Platyopuntia*) germinate successfully only when chewed and digested for a long time. During the Pleistocene period, this abrasion was provided by the digestive system of large ungulates, such as gomphotheres or giant ground sloths. At the

end of the last glaciation some 15 000 years ago, however, much of that Pleistocene megafauna went extinct — a fate that humans likely contributed to (Alroy 2001, Brook and Bowman 2004). Loss of that fauna resulted in the loss of seed dispersal and regeneration mechanisms for a number of plant species. Desert plant species with anachronic seed dispersal have merely survived for the last millennia through vegetative growth and accidental abrasion of seeds in the deserts' sand and gravel, in the absence of their effective seed dispersers. Not surprisingly, when humans reintroduced ungulates — cattle — into the New World some five centuries ago, the population of many of these plant species rebounded to large numbers.

Humans continue to affect desert ecology, at times fundamentally. Being areas of such low productivity, deserts can be easily degraded — even irreparably — by the increasing intensity of human land and resource use. Desert soils, which are of generally limited profundity and high fragility (see Box 1.1), are highly susceptible to compaction, erosion, and salinization when exploited for agricultural, industrial, or recreational purposes. Invasive non-native plants, whether introduced intentionally (such as in the case of the planting of grasses for livestock forage, which have the effect of introducing a grass-fire cycle to an ecosystem that has no natural fire regime) or not (such as the case of the invasion of *Tamarix ramosissima* in Nearctic deserts, which can substantially alter desert hydrological regimes), can have cascading effects on ecosystem function and native species viability in deserts. Human industry in and beyond deserts alters not only desert weather patterns via anthropogenic climate change, but also desert nutrient cycling via atmospheric deposition. Paradoxically, fertilization of deserts through increased deposition of nutrients like nitrogen can favour the invasive dispersal of non-native species and reduce native diversity. Whether through direct or indirect pathways, humans clearly have a hand in determining the future course of desert evolution.

CONCLUDING REMARKS

To the untrained eye, deserts look barren, especially during dry periods. However, because of their evolution in relative geographic isolation, most deserts of the world are rich in rare and endemic species, and are hence highly vulnerable to biological extinction and environmental degradation. In spite of their remarkable convergence in adaptation, deserts are different in their origin and their evolutionary history. Their incredible variation of the world's deserts in rainfall patterns, continentality, temperature regime, and evolutionary history have all contributed not only to their biological uniqueness, but also to their wondrous wealth of life-forms and adaptations, from some of the shortest-lived ephemeral plants, to some of the longest-lived giant cacti; from seed-eating rodents that do not need water to survive and depend on their burrows to regulate their metabolism almost as if the burrow was an extended part of their body, to amazing pollinators like nectar-feeding bats that migrate thousands of miles following the flowering seasons (Davis 1998). This adaptive diversity — what Darwin, strongly influenced by deserts himself, called "forms most beautiful and most wonderful" — is what makes deserts unique. In the hot deserts we may find giant cacti and trees with mammoth fleshy stems coexisting with some of the toughest hardwoods; ground-creeping succulents side by side with fog-harvesting rosettes, incredibly fast-growing annuals together with the hardiest drought-resistant perennials; shrubs of enticing odours with some of the nastiest, spiniest plants ever. Very few parts of the earth contain a richer collection of natural adaptations.

The fragmented evolutionary history of the deserts of the world has been the driving force of their biological rarity, of adaptation to local conditions, and of specialization to isolated environments. After millions of years in isolation, the forces of evolution and fragmentation have yielded unique life-forms in each desert, strangely-shaped desert plants and extraordinary animals. The world's deserts are biological and cultural islands, lands of fantasy and adventure, habitats of surprising, often bizarre growth-forms, and territories of immense natural beauty.

REFERENCES

Allan, J.A., Warren, A., Tolba, M., and Allan, T. (1993). *Deserts: The Encroaching Wilderness (A World Conservation Atlas)*. Oxford University Press, Oxford

Alroy, J. (2001). A multispecies overkill simulation of the end-Pleistocene megafaunal mass extinction. *Science* 292: 1893–1896

Axelrod, D.I. (1950). Evolution of desert vegetation in western North America. *Publications of the Carnegie Institute of Washington* 590: 215–306

Brook, B.W., and. Bowman, D.M.J.S. (2004). The uncertain blitzkrieg of Pleistocene megafauna. *Journal of Biogeography* 31: 517–523

Chesson, P., Gebauer, R.L.E., Schwinning, S., Huntly, N., Wiegand, K., Ernest, M.S.K, Sher, A., Novoplansky, A., and Weltzin, J.F. (2004). Resource pulses, species interactions, and diversity maintenance in arid and semi-arid environments. *Oecologia* 141: 236–253

Cloudsley-Thompson, J.L. (1996). *Biotic Interactions in Arid Lands*. Springer, Berlin–Heidelberg

Cooke, R.U., Warren, A., and Goudie, A.S. (1992). *Desert Geomorphology*. University College Press, London

Davis, S.J.M. (2005). Why domesticate food animals? Some zoo-archaeological evidence from the Levant. *Journal of Archaeological Science* 32(9): 1408–1416

Davis, W. (1998). *Shadows in the Sun: Travels to Landscapes of Spirit and Desire*. Island Press, Washington D.C.

Dimmit, M.A. (2000). Biomes and communities of the Sonoran Desert Region. In *A Natural History of the Sonoran Desert* (eds. S.J. Phillips and P.W. Comus) pp. 3–18. Arizona-Sonora Desert Museum and University of California Press, Tucson

Ehrlich, P.R., Dobkin, D.S., and Wheye, D. (1988). *The Birder's Handbook*. Simon and Schuster, New York

Ezcurra, E., Montaña, C., and Arizaga, S. (1991). Architecture, light interception, and distribution of *Larrea* species in the Monte Desert, Argentina. *Ecology* 72(1): 23–34

FAO (2002). *Pastoralism in the New Millennium*. FAO Animal Production and Health Papers no. 150. Food and Agriculture Organization of the United Nations, Rome

FAO (2004). *Carbon Sequestration in Dryland Soils*. World Soils Resources Reports no. 102. Food and Agriculture Organization of the United Nations, Rome

GLOBIO (2005). *Global Methodology for Mapping Human Impacts on the Biosphere*. http://www.globio.info/region/world/ [Accessed 19 April 2006]

Goudie, A., and Wilkinson, J. (1977). *The Warm Desert Environment*. Cambridge University Press, Cambridge

Holmgren, M., Scheffer, M., Ezcurra, E., Gutiérrez, J.R., and Mohren, G.M.J. (2001). El Niño effects on the dynamics of terrestrial ecosystems. *Trends in Ecology & Evolution* 16(2): 59–112

Loik, M.E., Breshears, D.D., Lauenroth, W.K., and Belnap, J. (2004). A multi-scale perspective of water pulses in dryland ecosystems: climatology and ecohydrology of the western USA. *Oecologia* 141: 269–281

Louw, G.N., and Seely, M.K. (1982). *Ecology of Desert Organisms*. Longman, London

Mabbutt, J.A. (1977). *Desert Landforms*. MIT Press, Cambridge, Massachusetts

Mares, M.A. (1980). Convergent evolution among desert rodents: a global perspective. *Bulletin of the Carnegie Museum of Natural History* 16: 1–51

Martorell, C., and Ezcurra, E. (2002). Rosette scrub occurrence and fog availability in arid mountains of Mexico. *Journal of Vegetation Science* 13: 651–662

McGinnies, W.G., Goldman, B.J., and Paylore, P. (eds.) (1977). *Deserts of the World*. University of Arizona Press, Tucson

Mittermeier, R.A., Myers, N., Robles-Gil, P. and Goettsch-Mittermeier, C. (1999). *Hotspots: Earth's Richest and Most Endangered Terrestrial Ecoregions*. Agrupación Sierra Madre, Mexico, and Conservation International, Washington, D.C.

Morton, R.R. (1979). Diversity of desert-dwelling mammals: a comparison of Australia and North America. *Journal of Mammalogy* 60: 253–264

Mulongoy, K.J., and Chape, S. (2004). *Protected Areas and Biodiversity: An Overview of Key Issues*. UNEP-WCMC Biodiversity Series No. 21, Nairobi, Kenya

NASA (2004). "Blue Marble" website http://earthobservatory.nasa.gov/

Olson, D.M., Dinerstein, E., Wikramanayake, E.D., Burgess, N.D., Powell, G.V.N., Underwood, E.C., D'Amico, J.A., Itoua, I., Strand, H.E., Morrison, J.C., Loucks, C.J., Allnutt, T.F., Ricketts, T.H., Kura, Y., Lamoreux, J.F., Wettengel, W.W., Hedao, P., and Kassem, K.R. (2001). Terrestrial ecoregions of the world: a new map of life on Earth. *BioScience* 51(11): 933–938 (maps available online at: http://www.worldwildlife.org/science/ecoregions/terrestrial.cfm)

Pianka, E.R. (1986). *Ecology and Natural History of Desert Lizards: Analysis of the Ecological Niche and Community Structure.* Princeton University Press, Princeton, N.J.

Pipes, R. (1998). *Hot Deserts (World Habitats).* Raintree, New York

Ricciuti, E.R. (1996). *Desert (Biomes of the World).* Benchmark Books, New York

Robichaux, R.H. (ed.) (1999). *Ecology of Sonoran Desert Plants and Plant Communities.* University of Arizona Press, Tucson

Safriel, U. (1990). Winter foraging behaviour of the Dune Lark in the Namib Desert, and the effect of prolonged drought on behaviour and population size. *Ostrich* 61: 76–80

Schmidt-Nielsen, K. (1964). *Desert Animals: Physiological Problems of Heat and Water.* Clarendon Press. Oxford

Sher, A.A., Goldberg, D.E., and Novoplansky, A. (2004). The effect of mean and variance in resource supply on survival of annuals from Mediterranean and desert environments. *Oecologia* 141: 353–362

Shmida, A. (1985). Biogeography of Desert Flora. In *Ecosystems of the World. Vol.12. Hot Deserts and Arid Shrublands* (eds. M. Evenari, I. Noy Meir and D. Goodall) pp. 23–75. Elsevier, Amsterdam

Thornthwaite, C.W. (1948). An approach toward a rational classification of climate. *Geographical Review* 38: 55–94

UNEP (1997). *World atlas of desertification* (2nd edition). United Nations Environmental Programme, Nairobi, Kenya

USGS (2005). *Global Land Cover Characteristics Data Base.* Earth Resources Observation and Science (EROS). United States Geological Survey. http://edcsns17.cr.usgs.gov/glcc/globdoc2_0.html [Accessed on 19 April 2006]

Zavala-Hurtado, J.A., Vite, F., and Ezcurra, E. (1998). Stem tilting and pseudocephalium orientation in *Cephalocereus columna-trajani* (Cactaceae): A functional interpretation. *Ecology* 79(1): 340–348.

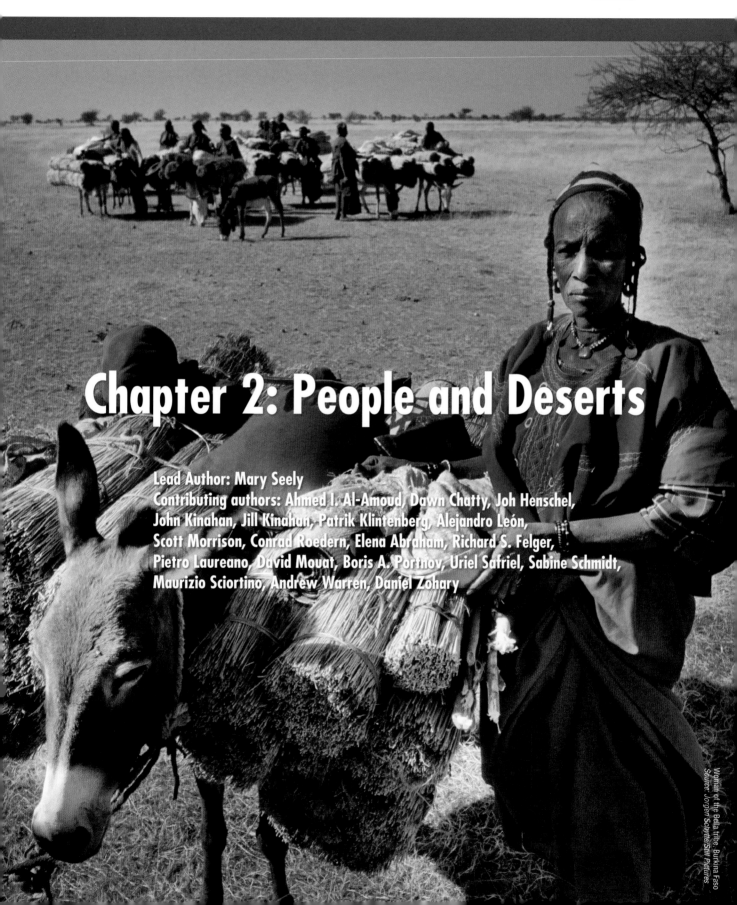

Chapter 2: People and Deserts

Lead Author: Mary Seely
Contributing authors: Ahmed I. Al-Amoud, Dawn Chatty, Joh Henschel,
John Kinahan, Jill Kinahan, Patrik Klintenberg, Alejandro León,
Scott Morrison, Conrad Roedern, Elena Abraham, Richard S. Felger,
Pietro Laureano, David Mouat, Boris A. Portnov, Uriel Safriel, Sabine Schmidt,
Maurizio Sciortino, Andrew Warren, Daniel Zohary

2

People in Deserts: An Overview

People have lived in and around deserts since time immemorial where their activities and use of natural resources have been, and are, governed by the basic parameters defining all deserts. Rainfall, essential for growth and reproduction of plants and animals, for grazing and for agriculture, is a central factor. High temperatures and strong winds also influence people's use of deserts. Adaptations of people to these elements are different, mainly in degree but not in kind, from those of other animals and of plants. People have relatively few morphological and physiological adaptations with a predominance of behavioural, cultural and technological adaptations. People have used a variety of approaches to live in deserts and continue unusual innovations.

Although limited, the same physiological principles governing, for example, heat exchange in animals and plants pertain to human thermoregulation (Louw and Seely 1982). Humans, unlike many large mammals, do not pant in response to heat. Instead, humans sweat profusely and no other animal sweats as efficiently to support evaporative cooling. People can produce up to 4.2 litres of sweat per hour if well acclimated. Surprisingly, many people under hot conditions undergo considerable dehydration before drinking to replace lost body fluids. Heat stress, from increased body temperatures exacerbated by dehydration, may range from temporary loss of consciousness to stoppage of sweating, circulatory failure and death. Overall, key factors supporting humans in deserts are an adequate supply of water and shelter from the sun's direct rays.

Meagre physiological adaptations of people to deserts are more than adequately augmented by behavioural, cultural and technological adaptations. People are able to thrive in deserts simply by modifying their micro-environment. These modifications range from using natural shelters, for example caves or shade trees, to using appropriate clothing, to construction of dwellings and use of air conditioning. Behavioural, cultural and technological adaptations have evolved to ensure adequate food, water and shelter. The result of these adaptations

has led to three major inter-related livelihoods: hunting and gathering, domestic livestock herding, and irrigated agriculture. While all these lifestyles are being practiced today, most have been extensively altered by modern technology.

Before describing traditional ways of resource use and management, brief consideration will be given to general aspects of people living in deserts. Protection from extreme heat and extreme cold is an important design consideration for desert clothing. Bedouin robes of light wool are considered to be an excellent compromise (Louw and Seely 1982). Evaporative water loss can be reduced by about one-third and heat gain by 55 per cent by wearing appropriate, loose-fitting clothing. Although white clothing will reflect solar radiation in the visible range, black or white clothes ensure the same body surface temperature. Nevertheless, two peoples living in deserts are known for wearing little clothing, the San people of southern Africa and the Aborigines of Australia (Biesele 1994, Bindon 1994).

Diet presents another aspect for consideration in hot deserts although basic requirements for high-quality protein, vitamins, minerals and sufficient energy naturally apply (Louw and Seely 1982). Adequate water intake is of primary importance and, contrary to popular opinion, the normal amount of salt used for flavouring meals is sufficient. Very high protein intakes are undesirable. If present in sufficient quantities, the traditional diet of West Asia, based on low-protein cereal grains and protein-rich leguminous seeds and featuring tea and coffee while excluding alcohol, fulfils most theoretical criteria for an appropriate diet in deserts.

Heat and aridity also are important in terms of housing. The physical principles governing the design of permanent desert dwellings are well-known. Thick walls and small windows protect from the day's heat but do not allow for cool air circulation in the often still night hours. In many areas, this leads to people sleeping outdoors or on the roof. Strong winds are also a consideration. These winds go by many names in different parts of the globe: the *Santa Ana* in California, the *föhn* in the Swiss Alps and the *zonda* in Argentina. People, other animals and plants must deal with the increased evaporation and very low humidity of these winds. Moreover, solid edifices are not available to many desert dwellers, and other adaptations, for example low tents or transportable Mongolian *ger*, suffice while addressing the frequent occurrence of strong winds in addition to heat and aridity (for example, Flegg 1993).

This chapter will examine past and present livelihoods of people living in deserts, and their ever-changing relationships to available natural resources in these lands of scarce and unreliable rainfall, abundant sunshine, high temperatures and strong winds.

Traditional Desert Dwellers: Resource Use and Management

HUNTER-GATHERERS

Early inhabitants of deserts, and all other environments, used resources in a way that is now described as hunting and gathering. In deserts it would have meant having the essential knowledge and being well-attuned to variable rainfall and the resultant growth patterns and behaviour of plants and animals, as well as to replenishment of ephemeral water sources. Certainly as far back as when *Homo erectus* occupied dry areas, it is thought that they used deserts on an intermittent basis when productivity was high (Shackley 1980). Interpretation of evidence from the central Namib Desert suggests that resource use was not simply a system of seasonal mobility, implying an almost random form of density-independent use of ephemeral resources (Kinahan 2005). Instead, an equilibrial or density-dependent system making use of key resource locations with reliable water during the dry season, within a wider area of ephemeral resources, would provide a better explanation.

A key resource essential for most groups of hunter-gatherers living in deserts is the presence of at least one tree-borne fruit that serves as a staple and is capable of long storage (Pailes 1999). This would be combined with grains, beans, roots and fruits, supplemented by small amounts of animal protein. In North America, acorns, pinyon pine nuts and mesquite beans fulfil this tree-fruit niche. In southern Africa, the mongongo nut and !nara seeds provide an equivalent. These resources could be harvested

and consumed on site, but served an increasingly crucial role as baskets or pottery for transport and storage became available. They would have provided sustenance for traders and even served as trade goods as such relationships developed.

On the coastal deserts, particularly important on the west coast of the southern deserts (for example, Smith and Hesse 2005), use of marine resources would have constituted a large part of the diet, at least seasonally (for example, Kinahan 2000; and see Box 2.1). Prolonged occupation of coastal areas entirely surrounded by dunes but with fresh water seepage indicates the importance of marine resources for early hunter-gatherers (Shackley 1983). This rich diet may have relieved the necessity for a tree-borne fruit as a staple although in areas such as the Namibian coast, the !nara fruit, growing in the coastal dunes and ephemeral water courses, would have provided the necessary component (Henschel and others 2004).

Although hunter-gatherers occupied deserts for millennia, evidence for their livelihoods comes from archaeological (Figure 2.1) and recent observations

Figure 2.1: Cave paintings, Baja California

The cave paintings of Baja California record in great detail the activities of early desert hunter-gatherers. Such paintings, with small local differences, can be found in many places in African and Asian deserts as well.
Source: Patricio Robles-Gil

Those who do not know the Sonoran Desert may find it inhospitable, but the people who lived here had vast knowledge of the land and its plants and animals, and their ways of life were rich and diverse. Plant and animal resources across the desert are as varied as the region itself (Hodgson 2001; Felger 2006; Nabhan 1985; Rea 1997). Depending on where one draws the limits of the desert, its vascular plant flora consists of about 2 500 species in an area of about 300 000 km². People had knowledge and names for most of the plants and visible animals that are today known as species. In fact, like most people with close ties to the land, their folk classification often approximates modern concepts of genera and species. About 20 per cent of the plant species probably were used for medicinal purposes and about 10 per cent for food   although major staples were derived from a much smaller number (Felger and Nabhan 1978). Hence, across the entire Sonoran Desert (as defined by Shreve 1951), about 375 species of plants have been used for food.

People who lived along the thousands of kilometers of coastal desert in northwestern Mexico had easy access to a wealth of seafood the world may never again know. The Comcáac (Seris) of Sonora and the culturally extinct people of Baja California made use of a cornucopia of sea turtles, fishes, and molluscs, as well as the usual terrestrial animals and plants. The Seris also harvested eelgrass (*Zostera marina*) as one of their staples—the only known case of people using a grain from the sea as a major food (Felger and Moser 1985). Other peoples visited the shores of the Gulf of California for salt gathering as well as for seafood. Edible molluscs were often carried inland; especially Venus clams (*Chione* spp.), which could be kept alive and fresh for several days (Marlett 2005). Such shells litter ancient trails and campsites, and sustained an extensive trade in ornamental shells from the Gulf of California across the whole region and beyond (Felger 2006; Haury 1975).

The Sonoran Desert is home to peoples from several extremely different language groups. Within the major groups, such as the Uto-Aztecan Piman-speakers and the Hokan/Yuman-speakers and the linguistically-isolated Seris, there were dialects and myriad geographic and cultural differences. Despite this diversity, people of the Sonoran Desert did not live in isolation. Information and goods flowed among neighboring and even distant peoples (Ford 1983). People traveled and traded, talked to each other, and sometimes intermarried; they fought and took captives and adopted them into their societies.

Desert peoples such as the Seris, the Cochimí and others of Baja California, and the Hia C'ed O'odham (also called Sand People, Pinacate people or Sand "Papagos") of Arizona and Sonora lived in the driest regions of the desert, and obtained their food resources from what we call "hunting and gathering." Some Hia C'ed O'odham people, however, also had access to desert oases, where they practised agriculture and established more permanent residences. These desert peoples had very small populations, mainly because their water resources were extremely limited. Although outsiders considered them nomadic or semi-nomadic, they moved their residences within well-prescribed geographic limits according to schedules that varied with the vagaries of rainfall. They relocated to take advantage of different food resources but undoubtedly also for aesthetic reasons and, most critically, because of scarce water. Every water place was known in detail, and each had a name (Broyles and others 2006a). I once located a legendary waterhole in the Pinacate region of northwestern Sonora by following an narrow, ancient trail.

Seri people—just like many other people—like living or camping along the beach. They travelled to islands and hunted at sea using reed boats, or *balsas*, made of interwoven bundles of reedgrass (*Phragmites australis*) bound together with cordage fashioned from the fibre of mesquite roots (*Prosopis glandulosa*). They would carry their balsas to the top of high beach dunes to catch the sea breeze and have an easy lookout. When they changed to much heavier wooden boats in the twentieth century, they had to move their camps down to the shore.

People such as the River Pimas in southern Arizona and other O'odham groups (Piman-speakers), living farther inland along rivers or in regions of higher rainfall, had sufficient water for irrigation. Rich agricultural traditions existed along the several rivers that crossed the desert—the Colorado and its tributaries—as well as in many regions in the interior desert but on a greatly reduced scale. Some prehistoric Hohokam people in southern Arizona deserts built complex irrigation canals to transport river water to their fields. Their agriculture sustained substantially greater populations than the hunting and gathering livelihoods of non-agricultural people but ultimately fueled cultural collapse when the population exceeded the region's carrying capacity (Diamond 2005). The prehistoric Trincheras people of northern Sonora constructed extensive rock-walled terraces on desert mountains (McGuire and Schiffer 1982). Most of the terraces were on north-facing slopes and probably were used for the cultivation of century plants (*Agave* spp.). Elsewhere, such as desert bajadas northwest of Tucson, people grew agaves on artificial rock piles or "rock-mulch" constructs (Fish and others 1992). There is considerable evidence that prehistoric people made use of selected agave cultigens or domesticated land-race varieties (Hodgson 2001). Agaves were harvested throughout the region. When mature, the plants concentrate carbohydrates just before producing their life-ending inflorescences, and at this stage they are harvested and pit-roasted to produce large, sweet vegetable roasts (Gentry 1982).

People such as the Tohono O'odham in northern Sonora and southern Arizona diverted run-off water from sporadic summer monsoon rains onto their fields, which were widely scattered to increase the chance of getting water from localized and unpredictable scattered thunderstorm rains. The agricultural mainstays were the usual Native American trinity of maize (corn), beans, and squash. In the arid and semi-arid regions of southwestern North America, teparies (*Phaseolus acutifolius*) were the most common beans and were probably domesticated from local, indigenous wild teparies. The major edible squash was an indigenous cushaw squash (*Cucurbita argyrosperma* var. *callicarpa*). Although these crops were similar to the agricultural staples of the great Mesoamerican cultures, the desert people developed their own, highly diverse, desert-adapted varieties or land races. Their maize varieties were selected to mature especially fast because of the desert's short summer monsoon season (Fish 2004).

A trinity of legume trees was especially important to most of the Sonoran Desert peoples: mesquites (honey mesquite, or *Prosopis glandulosa*, and the closely related velvet mesquite, or *P. velutina*), palo verdes (mostly *Parkinsonia microphylla*), and ironwood (*Olneya tesota;* Bean and Saubel 1972, Felger 2006). All three produce large crops of edible pods or seeds in early summer. In the summer-rainfall parts of the desert, the giant cacti (for example, saguaro, or *Carnegiea gigantea*, and cardón, or *Pachycereus pringlei*) provided seeds rich in protein and oil. The large, succulent and sweet fruits of columnar cacti, especially saguaro and organ pipe (*Stenocereus thurberi*), were harvested not only for food but also for making wine. Coastal-dwelling Seris spent the full moon of May going up and down the coast drinking cactus wine and partying. It was a good time of year to enjoy a pleasant sea breeze while sitting in the shade of a ramada (an open shelter). In early summer as the saguaro fruit ripened, the desert O'odham held wine-harvest ceremonies to invoke the summer monsoon.

Seri woman harvesting wild wolfberries (*Lycium fremontii*) in the coast of the Sonoran Desert, by the Sea of Cortés
Source: Patricio Robles-Gil

Both plant and animal foods were prepared fresh and often also dried and stored for future use. A wide variety of vegetables, greens, fruits, and seeds were dried whole, sliced, ground, or made into cakes and stored, as were many kinds of animal foods such as sphinx moth caterpillars (*Hyles lineata*), shrimp, fishes, sea turtles, deer (*Odocoileus hemionus* and *O. virginianus*), and desert bighorn (*Ovis canadensis mexicana*). Many desert seeds are digestible only when the seed coat is broken, as when the seeds are ground on a grinding stone (*metate*). Most kinds of seeds and many other plant foods such as mesquite pods were parched, toasted, or dried and ground into flour, then boiled or steeped in water to be consumed as *atole* (Bean and Saubel 1972; Castetter and Bell 1942, 1951). No satisfactory term for *atole* seems to be available in English, the nearest approximations being "gruel", "mush", or "porridge". It is nearly impossible to avoid sand and gravel when preparing food on a grinding stone that rests on the ground or on a dirt floor. I once tried bread from eelgrass flour prepared on a metate; it tasted good, but the sand was not pleasant. If we had consumed the flour as *atole*, the sand would have been left at the bottom of the pot.

Dogs were the only domesticated animals but were not eaten (Rea 1981). Hawks and eagles were sometimes kept in captivity for their feathers. Some people kept and raised macaws, likewise for their feathers (Rea 1983). Old World-domesticated animals and plants were first brought to the region with the Coronado *entrada* to New Mexico in 1540, and in the same year to the lower Colorado and Gila River region by the associated Alarcón expedition (Flint and Flint 2005). Few if any of Alarcón's introductions survived, apparently, although some of Coronado's did. Padre Kino was the first and most important agricultural extension agent for the Sonoran Desert. He brought cattle, goats, horses, and sheep, as well as wheat and many other Old World plants in the late seventeenth century. Juan Mateo Manje, soldier and close associate of Kino, reported that in 1696 at the Dolores mission in Sonora there were "fruit trees of Castile, grapevines, peaches, pomegranates, fig trees, pear trees, and all kinds of garden produce," all brought by Kino (Burrus 1971:95). Other Spanish colonial introductions were cowpeas, muskmelons, pumpkins, various squashes, and watermelons. The major indigenous Sonoran Desert agricultural crops are all warm- or hot-season, frost-sensitive crops. Thus, wheat was especially appreciated because it is a frost-resistant winter crop.

Although Sonoran Desert peoples harvested hundreds of species of plants, a few dozen formed the primary resources. One unique major wild crop was the grain of *nipa*, a salt grass (*Distichlis palmeri*), harvested by the Cocopahs at the delta of the Colorado River (Castetter and Bell 1951). *Nipa* is a word derived from the Cocopah name for this grass, which is wholly endemic to the intertidal regions of the upper Gulf of California. It thrives on pure seawater as well as brackish water, producing large yields of a grain about the size of wheat. It is a strong candidate for a major global food crop and could become this desert's greatest gift to the world (Felger 2006).

Native Americans used a cornucopia of Sonoran Desert food resource species, but do not forget that this is desert country. Extended drought could mean hardship (Clotts 1917; McGuire and Schiffer 1982). No matter how rich the resources, drinking water was an absolute necessity. Camps and settlements or villages and towns had to have access to waterholes and oases or one of the rivers. The desert people moved seasonal residences for a number of reasons, such as to follow their favorite harvests, but especially when their meagre water supplies were threatened, when a place became polluted with waste, or perhaps just to take advantage of more pleasant conditions. Drought in the distant upper reaches of the Colorado River could spell trouble for the Cocopahs and other downstream desert peoples (Castetter and Bell 1951). There certainly must have been drastic changes for the lower Colorado River people when the river changed course to empty into the former Lake Cahuilla (the remnant lake is the Salton Sea) and the delta went dry or the flow was greatly reduced (Cleland and others 2000).

A desert is defined not only by low rainfall but also by unpredictability (Ezcurra and Rodrígues 1986). Even coastal, maritime desert peoples such as the Seris of Sonora and Cochimís of the Baja California peninsula would be affected by the vagaries of weather and sea conditions. The occasional sustained winds or red tide meant curtailment of resources from the sea. Extended drought or late frosts in inland desert regions could damage essential wild crops such as palo verde or mesquite pods. Desert peoples had to know how to live on flexible schedules but still sometimes suffered from shortages (Broyles and others 2006b). Summer rains were eagerly awaited. For Sonoran Desert peoples, the new year began when the fruits and seeds of desert legume trees and giant cacti ripened, heralding the brief monsoon, for this is the time of greatest renewal of life in this part of the world.

Box author: Richard S. Felger

Figure 2.2: Topnaar wild harvest

Two Topnaar people collecting !nara melons (*Acanthosicyos horridus*) along the dry water course of the Kuiseb River in the Namib Desert. The pulp of these melons is preserved by the local people by boiling it in drums and spreading it out to dry.
Source: Mary Seely

from their remaining cultural descendents. The Topnaar of the Namib Desert today know uses for 81 plant species, although it is thought that some knowledge has been lost (van den Eynden and others 1992; Figure 2.2). The San, or bushmen, of the Kalahari have a repertoire of over 100 species of food plants and 55 animal species, a list also probably more extensive in the past (Biesele 1994).

Hunting and gathering may have been the only way of life known to people when they first occupied deserts, even intermittently. With the slow evolution of use of domestic crops and animals, the livelihoods of hunter-gatherers took on aspects of herding and agriculture in varying proportions. The remainder of this chapter refers to "mixed livelihoods" of people in deserts, although one way of making a living may predominate for a period of time, or in a particular area, or for certain parts of the population.

In the rapidly developing world of the 21st century, hunter-gatherers necessarily undertake mixed strategies of resource use while trying not to lose their rich resource base. The San (Barswara) of

the Central Kalahari are currently being relocated by the Botswana government out of what are perceived to be potential diamond mining areas. The San are an important element of most tourism experiences in southern Africa, and they have options for future development — although probably none represent their own preferred directions. In Namibia, a majority of the San are farm workers, mainly working for cattle-rich Herero peoples in the Kalahari sandveld (Gordon 2000), although they are also the nucleus of a very successful Community Conservancy focused on natural resource management and tourism (NACSO 2004). The future of the original hunter-gatherers of Australia, still very much focused on their traditional art and culture while adapting to modern ways, presents a similarly rich and complex transition (Bindon 1994, Kimber 2005). All modern day hunter-gatherers represent a repository of knowledge concerning life in desert landscapes, ready for interpretation and use that should not be ignored.

PASTORALISTS
Pastoralism refers to a livelihood approach that makes use of domesticated animals — for example, sheep, goats, cattle, camels — to provide a variety of products such as milk, skins, cash and occasionally, meat (Figure 2.3). Pastoralism evolved predominantly in Asian and African arid lands where most livestock were domesticated. Domestication is thought to have been undertaken by sedentary farmers rather than

Figure 2.3: Desert pastoralism

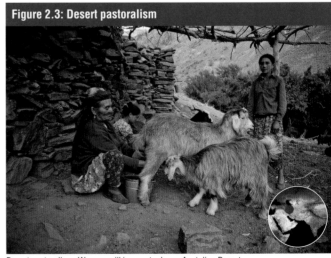

Desert pastoralism: Women milking goats, Irano-Anatolian Desert.
Source: Patricio Robles-Gil

hunters, as they would have had the capacity to corral animals for extended periods (Channell 1999a). Camels, the only livestock domesticated in hyper-arid deserts, are physiologically able to withstand desert conditions as they tolerate elevated body temperatures, are able to minimise water loss and reduce heat gain from the environment (Halpern 1999; Figure 2.4). They are able to tolerate water loss of more than 25 per cent of their body weight and can replace this within three minutes. They are without question well-suited to the desert environment and were responsible for most trans-desert trade before the advent of motor vehicles. Llama and alpaca are close relatives of Old World camels, and were domesticated to serve some of the same uses in the arid highlands of the Andes. Nevertheless, all mammals require water at frequent intervals, and in deserts this means herders must take livestock to temporary or permanent water holes, perennial or ephemeral streams and, in many instances, spend long hours lifting limited groundwater to the surface, often from great depths.

Other domestic animals are less well-adapted to deserts physiologically, but are nevertheless important for desert pastoralists. Sheep and goats represent the smaller, more tradable and expendable animals in southern African deserts, while cattle have greater associated prestige. Cattle (zebu in their humps), camels (in their humps) and sheep (in their tails) have concentrated fat deposits that store energy to carry them through times of limited pasture, but which do not hinder temperature regulation.

Based on different combinations of domestic animals, all forms of pastoralism incorporate an element of hunting and gathering, and many incorporate crop production as well. Most well-known are the nomads, with no permanent home base (Box 2.2). Entire families or groups move together with the herds. This is not a random movement, but usually follows fixed routes with careful scheduling based on rainfall and the presence of other herders (Flegg 1993, Pailes 1999, Smith 1994). Transhumance is the term given to livelihoods that include permanent villages and horticulture, augmented by seasonal movements of part of the group, often the men, with livestock to good grazing areas. This may

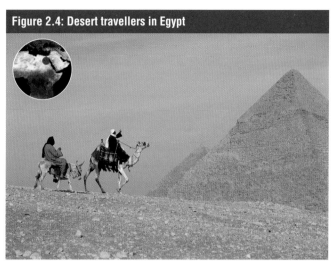

Figure 2.4: Desert travellers in Egypt

Desert travellers in Egypt approaching the pyramids on two traditional means of desert transport, a dromedary camel and a donkey.
Source: Patricio Robles-Gil

develop into a well-structured system where nearby grazing is protected and distant grazing used as conditions allow (Jacobsohn 1994, Pailes 1999). An important component of pastoralism is the presence of mobile merchants who, assuming many similar livelihood traits, support movement of products like salt, spices, grains, and necessities of life (Flegg 1993). The increasing complexity of interactions as pastoralism developed meant that individual families or small groups of hunter-gatherers changed to include more defined levels of organisation as an important component of living in and using ephemeral pastures and other resources of the desert landscape.

Mobility is a key to successful pastoralism in deserts as much as it is important for hunter-gatherers. Early herders followed rainfall and the variable grasslands that would appear in some areas in some years (Henschel and others 2005, Kinahan 1991). In some instances early herders harvested and ground natural grass grains as part of their resource base (Sandelowsky 1974). Nevertheless, there were distinct differences between strategies used to support differing livelihoods, as evidenced in the Namib Desert. While hunter-gatherers used reliable waterholes as periodic dry season gathering points, pastoralists with their herds spent dry periods in small dispersed camps. After good rains, pastoralists and their herds would aggregate wherever patchy rain provided good grazing (Kinahan 2005).

It is a surprise, in northern Niger, Chad or Sudan, to come across still-functioning nomadic economies. Is it such a good choice to live in a tiny tent hundreds of kilometres from the nearest shop? In fact, nomads could do far worse. Camels, sheep and goats fetch good prices, as settled people recognise when they invest large sums in livestock, and entrust them to nomads.

There have been for millennia, and they still are, purely sheep, goat, and cattle nomads in semi-arid areas, always many more than in the desert. The camel was the key that opened up the desert to the few. The Arabian camel (the dromedary) was adapted for nomadism in the 10th or 9th centuries BCE, at about the time of the domestication of the Bactrian (two-humped) camel in Central Asia. Both allowed much faster and further travel. Use of the camel spread throughout the dry parts of southwestern and Central Asia, northern Africa and the Horn of Africa. Extreme nomadism became both possible and necessary (in order to find sporadic grazing). Mobility had many other advantages: escape from oppressive laws and taxes, and contrarily, when these laws broke down, the ability to collect tribute from sedentary communities themselves.

Pastoral nomads cannot rely only on their herds; survival requires other things. To get these, nomads must trade their own products, sell their own labour (as for harvesting dates), or fish (for the market), if they live near the coast. Until 40 years ago they also transported travellers, pilgrims and goods (salt, drugs, slaves) on their camels. A few still do. Managing trucks, as they now must, needs similar skills in navigation and survival, and has taught new skills. Camels are still profitable for nomads in very arid areas, but sheep and goats, with higher rates of growth and reproduction and a better market, have replaced them in the less dry parts of the deserts.

When mobility was restricted by political boundaries, or the appropriation of land, or by severe drought, a small minority of nomads settled voluntarily. Some took up agriculture, but many transferred their innate adaptability to other theatres: finance, smuggling, or the military (Lancaster 1981, Chatty 1996).

How should planners now approach the nomadic realm? They have begun to act upon three realisations. The first is that nomads have minds of their own. The second is the large market for meat that nomads can produce from otherwise unproductive land. The third is the value of traditional, communal systems like *hema*; this Syrian approach has attracted praise in some quarters. After years of experimentation with Bedouin development, the government forbade cultivation in large grazing reserves, and developed a rangeland management programme for them. The results will need to be evaluated (Ngaido and others 2001). Another innovative approach is being tried with the Harasis nomads in Oman, who were asked what they felt about their future. Most wanted to keep livestock, for this was their main identity; some wanted to settle, but keep the nomadic option open; the new thing they most wanted was education (Chatty 1996).

Box author: Andrew Warren

The Himba, the "ochre people of the dry riverbeds", are well known and admired for their herding prowess, particularly by those living in higher rainfall areas (Jacobsohn 1994; Figure 2.5). With the advent of rain, they move westward to annual pastures of the desert margins, preserving more permanent desert springs and their surrounding vegetation for the late dry season. For such a system to work, social relations and organisation are very important to the entire group. This is encapsulated in their saying: "don't start your farming with livestock, start it with people". A system where inheritance through females relates to material wealth, while residential units and authority are inherited through males, serves to strengthen the needed social organisation. Contrary to many declining pastoral groups in arid southern Africa today, older people make sure that younger herders have full access to all available knowledge for successful use of the desert landscape.

For many centuries, pastoralism has been an important livelihood in the deserts of Asia, such as the Gobi (Box 2.3). Pastoralists manage the greatest proportion of desert lands, for subsistence or for profit, but alternatives ranging from hunting to tourism on desert pastures are emerging. On the other hand, increasing population, changing politics, enhanced educational opportunities and globalisation are all influencing pastoralists of the desert realm.

Today, pastoralists do not necessarily live in deserts but continue to herd their animals there to provide valuable products for urban consumption. Camels are being replaced by cattle with their better market value, and four-wheel drive vehicles provide transport (Smith 1994). It is a way of life that is rapidly changing, but nomads are still considered to be the most efficient producers of meat in deserts. Because of the large areas they use, they

Figure 2.5: Himba woman, Kaokoveld, Namibia

This young woman displays traditional jewellery with a white conch *ohumba* from the west coast as centrepiece. Her hair style indicates that she has recently completed the puberty ceremony and is now ready for marriage.
Source: Patricia Rojo

are profoundly affected by changes in politics and the environment, and their future is variable and unpredictable (Marx 1994).

IRRIGATED AGRICULTURE

Agriculture has been important to people in deserts since domestication of crops began. Rain-fed agriculture is less important in deserts than in higher rainfall areas because of the scarcity and unpredictability of rain; alternative systems began as attempts to reduce risks imposed by rainfall variation. Irrigated agriculture has evolved in different ways in different places based on different situations and crops available (Figure 2.6).

Large perennial rivers running through deserts, for example, the Nile, Tigris-Euphrates, Rio Grande and Colorado rivers, have supported desert people and their irrigation for a long time. Of particular importance on the Nile was the annual flooding and deposit of nutrient-rich silts on alluvial soils as well as prevention of salt accumulations (Dregne 1999a). All these benefits stopped once the Awsan High Dam was built in 1970 and alternative fertilisers and salt removal techniques are being

tested. The Nile Delta is shrinking for lack of silt deposits and the final solution has not been identified. The Colorado River supports some agriculture but is better known for supplying water and electricity to major urban centres in Arizona and California (Channell 1999b). Its delta has lost most of its water and hence its productivity. The Tigris and Euphrates basin has been occupied by people for millennia and is part of the Fertile Crescent with its early irrigation and urban developments (Dregne 1999b). The lower marsh has been recently drained and its contribution to agriculture eliminated. Silt in the river water and salinization of the irrigated land have been ongoing problems. Clogging by silt can be addressed by individual farmers for smaller canals, but peace and political stability are required for maintenance of large river systems of 300 km or longer.

Small-scale rainfall harvesting in deserts has been adapted to specific types of terrain, climate conditions and choice of crop (Lövenstein 1994). None of these approaches lend themselves to large scale, mechanised farming, but have provided abundant agricultural products for desert people. The terrace system was probably one of the

Figure 2.6: Modern desert irrigation

Mennonite irrigation farmer in the Chihuahuan Desert. Melons are commonly grown throughout the deserts of the world.
Source: Patricio Robles-Gil

Mongolia's Gobi Desert is an old cultural landscape, managed by nomadic livestock herders and sedentary cultures for millennia. Throughout the area, petroglyphs etched into dark rock faces illustrate human activity, their interaction with the environment, and the human view of wild and domestic animals over different periods of the last 2 000 years.

The Gobi Desert is a grand and diverse landscape from where vast expanses of steppe mountain ranges rise to rugged peaks, their slopes green with patches of juniper and forest remnants on more sheltered sites. Giant sand dunes contrast with deep gorges filled with ice even at the height of the desert summer. During the long winter and even more so in early spring, this desert steppe is exposed to high winds, and herders and their livestock seek shelter in their winter camps, often in lower mountain reaches or higher in quiet valleys, while alternating snow and sandstorms can continue for weeks. The Gobi is the coldest desert on earth, where the seasonal temperature difference within a year may reach 80°C.

A nomadic herder riding across Mongolia's Gobi Desert.
Source: Sabine Schmidt

Pastoral land-use practice has evolved over millennia to make optimal use of spatially and seasonally highly variable pasturelands in drylands that receive less than 200 mm of precipitation annually. Desert vegetation in turn has adapted to grazing by domestic livestock and by wild ungulates. Desert species important as livestock fodder plants have evolved buds close to the ground, allowing regeneration after grazing, and large areas of the Gobi every year are covered in a green carpet of a species of leek resistant to permanent grazing impacts. As long as mobility of livestock herds is maintained, desert steppe pastures are resilient.

Nomadic livestock herders in the Gobi used to undertake long distance migrations, moving seasonally north into the forest steppe zone. Centralized government and planning of administrative and economic units during Mongolia's socialist time, under Soviet dominance, brought an end to most long migrations. Pasture use coordination was directed within the state collectives and mobility was facilitated with state support. With the demise of the Soviet Union and collapse of the centrally-planned economy and government, a vacuum of institutions for grasslands management was left in rural Mongolia. For lack of other income opportunities, many households turned to subsistence herding, keeping private livestock on state-owned land. Many of those lacked traditional herding skills and had only small herds with not enough transport animals to move their camps. A cycle of poverty and land degradation set in, making already weakened livestock very vulnerable to cyclical winter disasters occurring in the region. A dramatic loss of 7 million livestock countrywide and a tragic rise in rural poverty occurred between 1999 and 2003.

The truth that institutions lie at the heart of livelihoods was felt by herders of the Gobi. Their pastures were degrading through a lack of coordination among them and through a lack of mobility. Throughout the area, pasturelands were becoming unusable for lack of water supply as wells were not maintained.

Triggered by the need to restore mobility, and supported by technical assistance, herder communities in the Gobi began to organize. Collective action, now voluntary, revived customary institutions for local natural resource management. Community norms for pasture use improved grazing resources and livestock. By working together, herding households could improve their preparedness for winter and natural disasters, add value to products and market them, access services and link with resource agencies. Livelihoods began to improve and pastoral community organization generated important lessons for poverty reduction in the country. A new adaptation was underway — customary institutions adapting to the new socio-economic and political landscape. The herder community organizations that emerged as institutions for natural resource management and conservation have become important actors in rural development and a driving force in a fledgling civil society.

New livelihood strategies are now being developed. Drawing on their rich cultural heritage, Gobi livestock herders now run their own Herders' Tourism Network, providing visitors with travel by horse and camel, and introducing them to the local culture and environment.

Box author: Sabine Schmidt

earliest irrigation systems involving a series of stone walls across a water course. With rain, the terraced fields would fill up and excess water cascaded onto fields below. Perhaps as a next step, the hillside conduit channel system was established based on narrow channels from neighbouring hill slopes. Micro-catchments, still in use today, have for thousands of years enhanced water application

"Oasis" has become a metaphor for refuge, and many oases in the desert are just this: welcome shade and security in a blisteringly hot and dangerous environment. The metaphor has had long enough to develop, because oases have a long history, and with care, they will have a long future. Some oases are visited only occasionally, for planting new trees, or for harvesting a few dates; some support a few hundred families in permanent homes; the oases around the old city of Bukhara, in Uzbekistan, support 1.2 million people, on 230 000 ha of irrigated land; those along the Tarim River in western China or the Nile in Egypt, support many millions of people. Huge cities, like Cairo, Damascus, Baghdad and Urumqi, depend largely on the production of oases.

Some of the ancient hydraulic systems that harvest water onto oases still astonish. The *qanat, foggara, karez* or *falaj* system leads water, from deep in an alluvial fan in a mountain basin, down a gently sloping tunnel to an oasis. The water is found with a well, sometimes hundreds of meters deep, and the tunnel from it is marked on the surface by a line of maybe hundreds of other wells for ventilating the well

Shaduf used by Topnaar people in the Kuiseb River, Namib Desert, to obtain domestic water and water for their livestock before solar pumps were introduced in the 1980s. The alluvial aquifer depth ranges to about 8m below the surface.
Source: Mary Seely

diggers and getting rid of their spoil. Other oases are fed by springs or mountain streams which are led into channels; some hacked out of the sides of gorges, others taken over small aqueducts. In northern Chile, pre-Columbian channels watered flights of irrigated terraces, covering hundreds if not thousands of hectares, now mostly abandoned. Yet other oases depend on water diverted from large rivers, at temporary or semi-permanent weirs, into intricate canal systems, as along the great rivers of early Mesopotamia, the Indian subcontinent (from the Harappan to the Moghul periods), Kazakhstan and western China. The Great Dam of Ma'rib in Yemen functioned for over a thousand years, and once watered 9 600 ha. It succeeded partly because it diverted rather than stored water, and therefore was not silted up, like many modern dams (Brunner 2000). Deep in the desert, far from rivers, as in much of the Sahara, oases depend on wells. If the wells are shallow, the water can be lifted to the surface by cantilevered bucket systems, or *shadufs*, by Archimedes screws, or by water wheels powered by oxen. All these systems (even at Ma'rib) are vulnerable. *Qanats* may collapse, and if specialist *qanat* engineers are not on hand, the dependent oases wither away. Rivers can change their courses, and isolate the off-take canals, or flood and sweep away the weirs that guide water into them. For these, and other reasons, many oasis systems have themselves collapsed. A newer threat, coming after the introduction of mechanical pumps and deep-drilled wells, is the heavy use of ground water, to the extent that the water-table retreats, as it has now done in many Saharan oases (Ebraheem and others 2004).

When water reaches a Muslim oasis, it is taken first past the mosque (where it is used for ablutions), and then to the gardens. In some Omani (and many other) oases, a committee controls the distribution of the water, so that no plot is dry for too long. A notice, pinned near the mosque, announces the next meeting of the committee. Most of these arrangements are cumbersome, and can be stultifying, so that many entrepreneurs today evade them by digging a well mechanically and installing a mechanical pump to feed a plot outside the old oasis. Communal systems that are less easily evaded are also needed to protect the encroachment of sand, as in Al Hasa in Saudi Arabia, in the Nefzaoua of southern Tunisia, or in the neighbouring El Souf region of Algeria.

Water brings nutrients to the fields, which are almost as important to sustainability as the water itself. In the San Pedro de Atacama oasis in northern Chile, where the river is particularly rich in silt, 37 500 million tons of material (a thickness of 180 cm) was deposited on irrigated fields between the mid-16th and the mid-20th century (Bork and others 2002). The water also brings salts, which accumulate in the soil as the water evaporates, and are a constant and common threat to sustainability. Moreover, studies of ancient systems, like that fed from the Ma'rib dam, or parts of Oman, show that some traditional systems used enough water to wash salts through the soil (Luedeling and others 2005), or irrigated very well-drained soils, as at Kharga in Egypt (Brookes 1989). If salts did build up, crops that were less susceptible to salinity, like the date palm, would gradually displace more vulnerable crops. But as the scale of water application increases, and unless drainage is managed, salinity can become a major threat to sustainability, as it is now in the oases of western China, and in Siwa in western Egypt, and as it once did in ancient Iraq. In oases that depend on well irrigation, salt is concentrated in the irrigated soil by evaporation, and may then seep back to the well water (Wang and others 2000).

Oases produce many types of crop, the most common being dates, other tree-crops like mangoes, vegetables, cereals like wheat and the more salt-tolerant barley, and fodder crops like alfalfa (and protein-rich quinoa in the Andes). Many local varieties of all these crops have been developed to suit local conditions (Moore and others 1994). Connoisseurs know the best oases for things like mangoes. Dates are now a major source of income in the north African (and some North American) oases. Other cash-crops, like early vegetables grown in plastic tunnels (as in Tunisia), or grapes for wine production (as in Argentina) are now profitable. The crops were once protected from pests by indigenous systems (Parrish 1995), but pesticides are now widely used (and overused). Harvesting was often achieved with temporary employment, as of nomads. Oases have a long future, but only if traditional wisdom about local conditions can be welded to new technology, and if new systems of environmental monitoring can be used to warn of threats to sustainability.

Box author: Andrew Warren

to small scale catchment areas. On a larger scale, diversions have been used on small and large, perennial and ephemeral rivers to channel water onto terraced fields on adjacent plains or even at a distance. Oasis agriculture has developed wherever water is available and has taken on many forms. A unique water harvesting system for oases is the *foggara*, which augment the water supply of isolated oases and small villages in places where it was not sufficient to encourage extensive settlement (Box 2.4 and Box 2.5).

With much ingenuity and hard work, different people in different areas supplied by an assortment of water sources have developed systems for crop production. Original savanna crops such as millet and sorghum have been adopted for desert use. Palms, tamarisk, acacia, and *Zizyphus* as well as tubers, fruits and cereals were all part of the Sahara array of crops when rains were more plentiful several millennia ago (Reader 1997). Palms,

however, are the only species known to have been domesticated directly in deserts (Box 2.6).

Hunting and gathering, pastoralism and irrigated agriculture all advanced, at different rates and different times, with different innovations. Baskets and pottery for transporting food and other goods allowed people to gather and carry more food and other materials. Deserts have always been a part of the global environment with desert peoples trading within deserts and with neighbouring cultures (see Chapter 3).

Population constraints within deserts, imposed by changing climates and agricultural developments, have caused wide fluctuations in the numbers of desert inhabitants (Reader 1997). With the expansion of technologies supporting people to live in deserts, the degree of fluctuation may be reduced. Nevertheless, institutions focused on resource management, as required for successful

Box 2.5: Harvesting underground water

The technique of underground water harvesting dates back thousands of years and has been adopted over very large areas stretching from China to Spain, and as far as Latin America, to mitigate the effects of arid climates.

Underground canals, called *qanat* in Iran, *foggara* in North Africa and Cyprus, *aflaj* in Oman, *karez* in Pakistan, *magara* in Jordan, *khottara* in Morocco and *madjirat* in Andalusia, may extend from a few hundred metres to several kilometres in length.

It is difficult to establish if these various systems derived from knowledge dissemination or from independent innovations in areas with the same physical characteristics. The existence of many ancient towns was based on these systems. One variation of this method of water production and the associated complex management procedures is efficiently used in the regions of Gourara and Touat in the Algerian Sahara desert with about one thousand *foggaras*, half of which are still working.

Foggaras consist of underground tunnels dug parallel to the ground surface. They do not reach the groundwater but drain off soil water, preventing lowering of the aquifer. The subsoil area for water supply acts like a big rocky sponge rather than an underground basin. *Foggaras* can be easily identified by the wells on the surface, recognized by their characteristic raised edge resulting from the excavation wastes. The wells are dug about 8 to 10 metres apart in order to guarantee proper ventilation during the underground digging; they are also used for maintenance work but are not used for extracting water.

Excavation of foggaras starts from the settlement site and follows the edges of the alluvial cones of the fossil wadi. Unlike a feeder canal, the foggaras do not convey water from springs or underground pools to the place where it is used. Instead, they tap micro-flows seeping through the rocks.

Foggaras may be supplied by three different processes:

1) Underground water flowing under the sands of an *erg* coming from distant rainfall on the highlands, for example, on the Saharan Atlas. Precipitation happened thousands of kilometres away, and it takes the micro-flows thousands of years to cover this distance under the sands of the *erg*, and to reach the oasis where the prehistoric rainfall is harvested.
2) Atmospheric supply from rainfall, which in these regions does not exceed 5-10 mm per year. Though it is quite a small amount of water, because of the enormous basin size, it can provide an oasis with a significant contribution.
3) Condensation of atmospheric water vapour may also contribute.

In contrast to our understanding of the *qanats* in Iran, quantitative assessments of the contribution of the three processes to foggara productivity and of the way the air condensation drainage tunnels work have not been elaborated.

Box author: Pietro Laureano and Maurizio Sciortino

Box 2.6: Plant domestication in desert environments

The climatically extreme desert biome is not an ideal location for practicing agriculture or for domesticating plants. In fact, in hot and dry desert areas, the shift from hunting and gathering to farming and herding took place relatively late. It depended, almost exclusively, upon two factors: (a) human introduction of crops that originated in milder and more humid biomes, such as steppes, grasslands, chaparrals, savannahs and park forests, and (b) the development of various types of irrigation systems. The only indigenous wild desert plant definitely domesticated in its native harsh environments appears to be the date palm (Zohary and Hopf 2000). Also, prospects for future domestication of true desert plants are far from bright. Medicinal plants as well as large ornamental species (such as succulent Chenopodiaceae, cacti or Euphorbias) seem to have better chances.

Date palms are the only fruit tree that were strictly domesticated in the desert.
Source: Daniel Zohary

The date palm, *Phoenix dactylifera,* is the best-studied example of successful domestication under extreme desert conditions (Wrigley 1995). Both cultivated varieties and wild forms of this fruit tree are adapted to extremely hot and dry climatic regimes that prevail in the West African, Arabian and Saharan deserts. Today, date palm plantations are a characteristic feature of sparse oases. Wild-growing forms of this palm (from which the crop could have been derived) also thrive in these harsh territories. Archaeo-botanical finds indicate that they are indigenous, at least in West Asia, including Arabia.

The date palm is a very productive fruit crop and a basic, staple food of local people. Ripe fruits of cultivated date palm varieties are packed with sugars (about 70 per cent of dry weight) and fruit yield is about 30-40 kg per tree under primitive farming and two to three times these amounts in modern plantations. It is extensively planted in Saharo-Arabian countries as well as in southern California and Arizona. The present annual world production is over 5 million tons, mostly consumed locally.

For effective pollination, fruit setting and fruit maturation, the crop and its wild progenitor require mild winters, intensely hot and fully rainless summers, very low relative humidity and permanent water supply (which can be quite brackish). Arab folklore has cleverly summed up the ecology of the date palm: "Its feet are in the water and its head is in the fire of the sky".

Closely associated with development of date palm horticulture was the Bronze Age domestication of the single humped dromedary camel, again probably in Arabia and/or other parts of West Asia. Similar to the date palm, the camel is a desert specialist perfectly adapted for survival in harsh, arid environments. Herding and rearing of camels added a supply of milk and meat to the date palm diet, as well as strong pack animals capable of traversing the hostile Saharo-Arabian deserts. This combination was largely responsible for opening these vast territories to human activity.

Box author: Daniel Zohary

hunting and gathering, nomadism, transhumance and oasis agriculture, will undoubtedly play a large, although altered, role in future desert development.

Modern Desert Dwellers: Resource Use and Management

People living in deserts today vary qualitatively and quantitatively in their development trajectories and their use and management of desert resources (Box 2.7). Some groups are more or less successfully continuing in their traditional ways, often against great odds, while adapting to developments taking place around them. An example would be the Harasis nomads in Oman: some are maintaining their livestock while others are settling down, but all want education to understand, if not completely partake of, "modern"

developments. Others, such as the Topnaars of the ephemeral Kuiseb River in Namibia, consciously maintain their traditional roots in villages along the river where old people oversee the livestock and bring up children not old enough to attend school (Henschel and others 2004). Irrigation farmers along the Nile continue farming while adapting to controlled river flow and its reduced silt load and increased salt deposition. Camel nomads of the Sahara continue long-distance trade and transport, but have shifted emphasis from camels and trade goods to motor vehicles and tourists. Similarly, a majority of bushmen (San) in Namibia are farm workers while others are members of a conservancy focused on wildlife conservation and tourism. Nevertheless, the San supplement their income using natural products from the veld and maintain their art and cultural traditions for themselves and for tourists. Because deserts

Mendoza, located in central-western Argentina on the eastern piedmont of the Andes, is a clear example of the opportunities offered, and the restrictions imposed, by deserts. A region of natural contrasts, it is also marked by deep social contrasts reflected in two Mendozas. One, rich and ostentatious in the irrigated oases, is built at the expense of the other: the non-irrigated desert, only inhabited by isolated producers of livestock, mostly goats (Abraham and Prieto 2000).

The 1 000 m elevation divides the territory into two halves, with mountains and piedmonts on the west, and lowlands on the east. The main perennial rivers, which feed from snow and glaciers in the Andes, form large alluvial fans on the plains, allowing development of irrigated "oases" that concentrate the productive, social and political life of the province. Waters of the Mendoza and Tunuyán rivers in the north, the Diamante and Atuel rivers in the centre, and the Malargüe river in the south, supply irrigation water for between 2.5 and 4 per cent of the province's area. Despite their limited extent (approximately 3 600 km^2), these oases support nearly 95 per cent of the population. Productive activity in the oases is structured as an agro-industrial model in a market economy, made possible by systematized irrigation and use of groundwater. Industrialization is mostly associated with produce from vine, fruit and vegetable crops (Roig and others 1991).

Water use is based on water law and on irrigation endowed with an active democratic participation of water users. Users vote for water supply managers who are responsible for distribution of water to each user and for water-use schedules, and who have their own funds. Advisors representing different rivers in the province and commissions at the provincial departments complete this administrative model.

Territorial imbalance appears in the development of irrigated oases to the detriment of areas lacking in irrigation water, and is crystallized in the contradictory phrases, *"culture of the oasis" – "culture of the desert"*. Competition for water use emerges as one of the primary environmental conflicts in the interaction between oasis and rain-fed areas. Rain-fed areas are characterized by low population, inefficient road systems, and dependence on equipment from distant urban centres. Extensive livestock and cattle-rearing predominate (Montaña and others 2005).

Economic globalization and its related integration processes impose new rules for development on the society and economy. The perspectives of provincial development are focused not only on consolidating access to international markets but also, and most importantly, on consolidating policies likely to result in higher territorial and social equity. When policies are formulated only for the oases, decisions are being made, by omission, about non-irrigated spaces, submitting them to a subordinate role. The myriad challenges include: developing systemic criteria for water use, (both surface and groundwater), rationalization of livestock rearing, production diversification, legalization of land tenure for the "occupiers" of the desert, territorial arrangements to guide urbanization processes, equity in the distribution of water resources, incorporation of appropriate technologies for water use and for the use of resources in general, as well as the recovery of traditional knowledge.

Box author: Elena María Abraham

harbour a number of diverse and seemingly exotic cultures living by a variety of lifestyles, they are inherently attractive to tourists from temperate climates. While some desert people continue their usual livelihoods and have "inadvertently" become tourist attractions, others are integrated directly into or are working for the tourism industry – with many levels in between.

Another culture whose presence has long been felt in deserts is that of mining. Wares of gold miners were carried across the Sahara to Europe by camel caravans for centuries. Salt mining produces another image associated with deserts and camels in northern Africa as part of a long-established, integrated trade network. The coasts of South America and southern Africa were densely populated for a few years while ships from Europe removed guano accumulated over millennia, or as in Namibia, established whaling stations for a few years (Kinahan 2000). Another type of mining has become almost synonymous with deserts over

the past several centuries, that of extracting oil and, more recently, uranium. The riches generated by these different types of mining supplied the income necessary to import missing resources from food to water and infrastructure. Large urban areas in deserts are now entirely dependent on imported resources, for example energy from oil for desalination of all domestic water and all other energy required by the city of Kuwait.

Several entirely different groups are taking advantage of the desert's vast landscape and unique scenery, dry air and almost continuous sunshine to relocate from less comfortable climates. With no traditional ties to these areas, retirees with adequate funds, those seeking a health cure, those seeking nearby recreational opportunities, or those just wishing to live in less arduous environments, are flocking to what were only lightly populated desert areas just a century before. Resorts, golf complexes and shopping megalopolises are also booming in desert

Most major population centres of the contemporary world are located outside deserts. At their beginning, many of these centres were established as small agricultural communities for which access to fresh water was essential (Portnov and Erell 1998). In contrast, urban development in deserts often generates sufficient economic outcomes to justify the considerable costs of importing fresh water (Portnov and Safriel 2004).

Historically, desert settlements have been scattered and sparse, and served as commercial and administrative centres, which sprang up around mines, transportation routes, and other local amenities (Saini 1980; Golany 1979; Issar 1999). Some were established as strategic outposts in response to various geo-political and security considerations (Portnov and Erell 1998). Today, desert towns and cities function as irrigation centres, garrisons, communications nodes and political, administrative and regional centres; they also may be focused on tourism, recreation, mining or other industries (Kates and others 1977).

There are four major factors contributing to a recent increase in the pace of desert urbanization:

Planned (bottom) and unplanned (top) development in a desert urban centre, Nouakchott, Mauritania.
Source: Stefanie M. Herrmann

- Relocation of territory-consuming enterprises, military and research installations from overpopulated core regions to peripheral desert areas;
- Mining and power engineering facilities, as resources are depleted in traditional mining centres and other less remote non-desert locations;
- Development of transport infrastructure which extends the commuting frontier of existing population centres into more remote desert hinterlands;
- Development of means of pumping fresh water over considerable distances from natural sources, and desalination technologies which have become more available and affordable (Portnov and Erell 1998).

Predicted global warming will also draw the desert frontier closer to many existing population centres, thus bringing more cities currently located at desert fringes closer to the desert; some may even become desert cities eventually. Thus, as a result of climate change, more non-desert people of today are likely to become desert inhabitants of tomorrow.

Concurrently, two major factors reduce the attractiveness of desert regions for newcomers — limited and undiversified employment opportunities and remoteness from major cities, which are major foci of employment, services and cultural life. However, if these drawbacks are mitigated, desert cities may exhibit impressive growth (such as in, for example, Tucson and Phoenix, Arizona), outranking even long-established non-desert communities. Environmental impacts of desert urbanization are not negligible. Compared to agriculture, however, urban development is compacts and economical on land use, thus leaving more spaces for environmentally-compatible desert uses, such as tourism and recreation (Portnov and Safriel 2004).

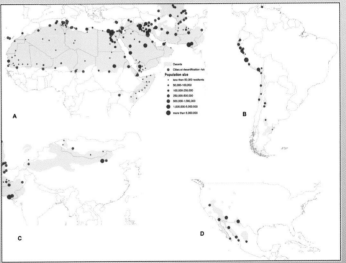

Desert cities of today (red dots) and cities located at desert fringes (blue dots), which are being at risk of becoming desert cities in the future due to anticipated climate change and resulting desert expansion. (A) Africa and West Asia; (B) South America; (C) Central and South Asia; and (D) North America.

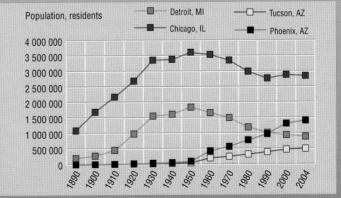

Comparative trends of population growth of selected cities of the United States
Source: U.S. Census Bureau

Box authors: Boris A. Portnov and Uriel Safriel

environments (Box 2.8 and Box 2.9). These new uses are developing simultaneously with other uses of the vast open spaces of the desert, such as military training grounds or, as in western China, resettlement areas to open up "new frontiers", release population pressure elsewhere, and provide a buffer against neighbouring states. While deserts are opening up for urbanisation and associated use, areas including cites and towns bordering deserts may inreasingly find themselves within deserts as global climate change takes place.

Resource use and management in desert areas for "modern" development focuses on two key resources, one of which is very scarce and one of which is highly abundant. By definition, water in deserts is limiting . It is usually brought from great distances (like in southern California), often disadvantaging the people from where it comes (Reisner 1986). It may require construction of large dams cutting off people in the lower reaches, as in the Colorado River. Or it may result in drawdown of hard-rock or alluvial groundwater aquifers altering local availability as well as livelihoods of distant populations (de Villiers 2000). Desalination of water for domestic use is increasingly considered and extensively used where energy is abundant. Appropriate use of water in deserts will have to be

Box 2.9: Las Vegas

Las Vegas, home to the fastest growing area of the United States, is situated in a valley, or down-dropped fault block, within the Basin and Range Physiographic Province. The valley is filled by recent alluvial and lake bed deposits from the surrounding mountains. The name, "Las Vegas", means "the meadows" in Spanish, a name that might confuse today's visitors and even residents as there are few meadows to be seen amidst the sprawling and endless stretches of subdivisions, casinos, malls, and other developments. Ironically, for an area that was known in the 19th and early 20th centuries for its abundant water resources, the Las Vegas area is one of the great consumers of water in the American West.

One reason for this water consumption lies in its arid climate (an average of approximately 100 mm of rainfall per year, combined with temperatures that regularly top 45°C in the summer and can reach 48°C), but the main reason is its spectacular growth and the need to appear well-watered to tourists. Las Vegas has been the fastest growing large metropolitan area in the United States, growing from approximately 340 000 people in 1972 to over 1.7 million by 2002. The 1972 and 2002 Landsat images illustrate the growth. The implications are tremendous. Certainly it is an economic boom to the various industries that provide for the growth. Most of the 35 million tourists who visit the area also probably see it as favourable. But growth also has costs. The meadows (the photo of nearby Ash Meadows illustrates what the area probably looked like prior to urbanization) are virtually no longer present. Many unique plant and animal species have vanished. Recreational activities for both residents and tourists mean increased pressure on a fragile desert ecosystem. One of the most severe threats comes from proposals to import water from lightly populated east central Nevada, home to just a few thousand people who live above a tremendous aquifer. A transfer of that water to Las Vegas could well bring an end to numerous seeps and springs by drawing down (lowering) the aquifer.

Ash Meadows is located about 150 kilometers from Las Vegas near Death Valley National Park. According to Dr. Don Sada, an internationally-known arid system aquatic ecologist, Ash Meadows represents a system that supports 24 endemic fish, plants and aquatic invertebrates. All of these fish and plants are listed as threatened or endangered by the U.S. Fish and Wildlife Service. This is the highest concentration of locally distributed endemic species known in the U.S. and the second highest known on the continent. Las Vegas, at one time, probably looked somewhat similar.
Source: Don Sada

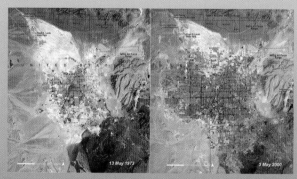

Landsat images of Southern Nevada including the metropolitan area of Las Vegas. In arid areas such as this, there is little natural vegetation cover, except for riparian zones. The Las Vegas Wash, located in the lower right center of each image, has dense vegetation and appears bright green. It can be seen emptying into an arm of Lake Mead. Las Vegas, itself, is largely green, indicating irrigated landscapes and golf courses.
Source: UNEP/GRID-Sioux Falls

Box author: David Mouat

re-evaluated in the decades to come, particularly as food production is an increasing competitive use for growing desert populations.

Energy is the second key resource essential for "modern" development and it is present in deserts, again almost by definition, in great abundance. To date, developers in deserts have largely ignored the abundant solar energy available and relied on increasingly expensive traditional sources of energy, like water, often brought from great distances, or otherwise on polluting the clear desert atmosphere – a main reason people come to the desert environment in the first place. Abundant solar energy could contribute to development not only of deserts but the entire globe (see Chapter 5).

ALTERNATIVE BENEFITS AND USES OF DESERTS

Deserts have not only supported and continue to support a variety of livelihoods; they have contributed extensively to global culture, traditional and modern (Figure 2.7). Three of the world's major religions had their origins in the deserts of West Asia (Mares 1999). Judaism, Christianity and Islam, the three "religions of the book", grew out of the profound religious experiences of desert cultures. All three religions are monotheistic, and today have enormous geopolitical influence extending far beyond their area of origin.

On a totally different level, today's culture and perspectives on deserts are greatly influenced by movies made in natural desert landscapes where the harsh habitat is usually portrayed as hostile to the presence of people (Ocampo 1999). A common theme of these films is isolation in a vast, arid landscape with its climatic extremes, scenic contrasts and limited supply of water. As a result, in the United States and to a lesser extent in Australia, the desert environment, combined with efforts to overcome its trials and tribulations, has become symbolic of national character.

Partly fuelled by landscapes and challenges depicted in movies, deserts have become favoured destinations for tourism and outdoor recreation. However, tourism, if it includes pilgrimage, is nothing new in deserts, and is a strong driver of change. Small-scale pilgrimage has a long history and is a common modern practice in deserts (Marx

Figure 2.7: Traditional cultures

A Seri woman performs an age-old ritual ceremony in the coast of Tiburón Island in the Sea of Cortés, Mexico
Source: Patricio Robles-Gil

1977), but by far the most important pilgrimage in the deserts today is the Hajj to Mecca and Medina. The Hajj now attracts some four million pilgrims, in just one month of each year. It is certain to achieve yet more change, not only in Saudi Arabia, where it already generates much income and fast urban growth, but also by bringing Muslims together from the largely Islamic deserts of the Old World.

In a different sphere, camping, hiking, fishing and hunting are all popular in deserts among those seeking sunshine, warm weather, unusual landscapes and interesting plants and animals. For the same reasons, however, and encouraged by sparse vegetation cover, off-road vehicle use is also very popular in deserts (Lacher 1999). Use of dune buggies, dirt bikes, quad bikes and ordinary 4x4 vehicles produces noise, disturbs wildlife and destroys the soil surface and vegetation. Disruption of soil surface can lead to increased wind and water erosion, loss of organic material and compaction of soil which reduces water infiltration. Archaeological sites are particularly prone to destruction. On the gypsum plains of the coastal Namib desert, the rich lichen cover is eliminated by a single passage of an off-road vehicle and tracks remain visible for decades if not centuries (Seely 2004). Gypsum plains and nearby dunes, including nesting sites of the endangered Damara Tern, are damaged every year during the "festive season" when hundreds of quad bikes descend on the small coastal desert resort of Swakopmund. The potential and ongoing impacts of off-road vehicles have led or are leading to increased management of public desert lands at many locations worldwide, although unclear jurisdictions and fear of revenue loss curtail many efforts (Figure 2.8).

The conservation of desert areas has had a chequered history and faces an unsure future. Deserts are often viewed as wastelands, uninteresting and useful for little more than perhaps prospecting and mining or military testing. In Namibia, the proclamation of the Namib-Naukluft Park had its origins in the colonial era when the Germans wanted to constrain the British to a small section of the coast at Walvis Bay. Decades of neglect followed as conservation efforts focused on areas where the expected big mammals of Africa are more common. Meanwhile, diamond, uranium and copper mines together with extensive prospecting and army activity left their indelible mark on the desert landscape. In the past several decades, the emphasis has shifted and Sossus Vlei, an ephemeral river terminating amongst the 300m high dunes of the Namib sand sea, is the second most visited tourism destination in the

country. Transboundary parks, being negotiated with Angola and South Africa and occupying most of Namibia's coast, could mean varying degrees of conservation for the entire coastal Namib desert.

In North America, Joshua Tree National Park, 80 per cent of which is designated as wilderness area, contains portions of the Mojave and the Colorado deserts. Three main vegetation zones support diverse fauna (Braun 1999). Established in 1936 as a national monument, it was declared a national park in 1994. Death Valley is perhaps the most famous desert park in the United States (Hulett and Charles 1999a), with its name alone evoking visions of desolation and harsh landscapes. Nevertheless, it is a popular tourism destination with its historical and archaeological connections, endemic fish species, extremes of aridity interspersed with occasional massive flooding, geological diversity and striking landscapes. Major protected areas have been proclaimed in Australia, Mongolia and Oman and other desert countries of the world. They may focus on the landscape and biota, for example Joshua Tree National Park, the cultural relevance of the desert area, for example Saint Catherine's Monastery in the Sinai or Ayers Rock in Australia, or its seasonal support to pastoralists, for example in the Gobi Desert. Conservation of desert landscapes is expected to increase and, with growing use of environmental assessments and general environmental awareness, to be in greater harmony with ongoing extractive and currently destructive uses.

Conservation of deserts has gone hand-in-hand with desert tourism. Desert movies, desert books and other awareness-raising media have contributed to the tourism drive. However, desert tourism can be seen on a continuum with desert recreation and the mix is not always a happy one. Nevertheless, tourism is growing and expected to be the main means of generating income in many desert areas of the world. In Namibia and South Africa, community-based tourism is a rapidly growing sector involving people who were formerly struggling to make a living from the arid landscape (NACSO 2004). The potential for tourism growth, in terms of quality of experience and number of attractions and people experiencing these options, is huge.

Figure 2.8: Desert vegetation patterns

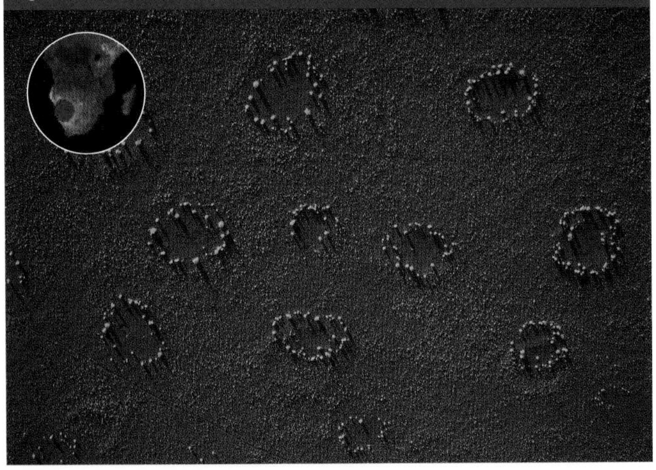

Vegetation in desert plains often forms surprising and fragile patterns that are easily disrupted by motor vehicles. These enigmatic "fairy circles" are common on the sandy inland edge of the Namib Desert, and are a popular tourist attraction.

Source: Patricia Rojo

Since the early days of the last century, if not before, desert research has held a special attraction for those who are interested in subjects ranging from geology to biology and from culture to religion all related to the extremes of desert environments. Some of the research efforts were small, one-person efforts, such as those of Felix Santschi who is identified as introducing to the scientific world sensory ecology based on his work with desert ants and ant navigation in Tunisia in the 1940s (Wehner 1990). Another was P.A. Buxton who first drew attention to the paradox of animal coloration in deserts and of so many black rather than the expected white beetles in the Palestinian desert (Buxton 1923). Large international research programmes, on the other hand, such as the International Biological Programme (IBP; 1966-1974) left a legacy in its multidisciplinary approach (Hulett and Charles 1999b).

Using a different, localised approach, Deep Canyon on the western edge of the Colorado Desert is associated with the University of California, Riverside located in the P.L. Boyd Deep Canyon Desert Research Center. It receives a variety of visiting scientists and students and, in addition to research, addresses conservation issues of the surrounding environment such as the fate of the fringe-toed lizard. Another desert centre established by one visionary biologist, the Gobabeb Training and Research Centre, is located in Namibia within the driest part of the coastal Namib Desert in the Namib-Naukluft Park.

People have lived in deserts for millennia, as hunter-gatherers, agriculturalists and pastoralists, and some people continue to do so today. But other people now live in urban developments situated in deserts, or enjoy deserts temporarily for tourism or recreation. Yet others are extracting profits from mining or other non-renewable resources. Deserts are a large and probably growing environment globally and their future will be best supported if it is based on a thorough understanding of their structure and function, and the influence of people's activities in the past, present and future.

REFERENCES

Abraham, E.M., and Prieto, M.R. (2000). Viticulture and desertification in Mendoza, Argentina. *Zentralblatt für Geologie und Paläontologie* 1(7/8): 1063–1078

Bean, L.J., and Saubel, K.S. (1972). *Temalpakh: Cahuilla Indian Knowledge and Usage of Plants*. Malki Museum, Banning, California.

Biesele, M. (1994). Bushmen of the Kalahari. In *Deserts* (ed. M. Seely) pp. 104–105. The Illustrated Library of the Earth, Weldon Owen, Sydney.

Bindon, P. (1994). People of the Australian Desert. In *Deserts* (ed. M. Seely) pp. 116–121. The Illustrated Library of the Earth, Weldon Owen, Sydney.

Bork, H.R., Bahr, J., Bork, H., Brombacher, M., Demhardt, I.J., Habeck, A., Mieth, A., and Tschochner, B. (2002). Development of the Oasis of San Pedro de Atacama, Chile. *Petermanns-Geographische-Mitteilungen* 146(5): 56–63

Braun, J.K. (1999). Joshua Tree National Park. In *Encyclopedia of Deserts* (ed. M.A. Mares) p. 316. University of Oklahoma Press, Norman.

Brookes, I.A. (1989). Above the salt-sediment accretion and irrigation agriculture in an Egyptian oasis. *Journal of Arid Environments* 17(3): 335–348

Broyles, B., Evans, L., Felger, R.S., and Nabhan, G.P. (2006a). Our Grand Desert: A Gazetteer. In *Dry Borders: Great Natural Reserves of the Sonoran Desert* (eds. R.S. Felger, B. Broyles). University of Utah Press, Salt Lake City.

Broyles, B., Rankin, A.G., and Felger, R.S. (2006b). Native Peoples of the Dry Borders. In *Dry Borders: Great Natural Reserves of the Sonoran Desert* (eds. R.S. Felger, B. Broyles). University of Utah Press, Salt Lake City.

Brunner, U. (2000). The Great Dam and the Sabean Oasis of Ma'rib. *Irrigation and Drainage Systems* 14(3): 167–182

Burrus, E.J. (1971). *Kino and Manje: Explorers of Sonora and Arizona*. Jesuit Historical Institute, Rome.

Buxton, P.A. (1923). *Animal Life in Deserts*. Edward Arnold and Co., London.

Castetter, E.F., and Bell, W.H. (1942). *Pima and Papago Indian Agriculture*. University of New Mexico Press, Albuquerque.

Castetter, E.F., and Bell, W.H. (1951). *Yuman Indian Agriculture*. University of New Mexico Press, Albuquerque.

Channell, R. (1999a). Domestic animals. In *Encyclopedia of Deserts* (ed. M.A. Mares) pp. 182–184. University of Oklahoma Press, Norman.

Channell, R. (1999b). Colorado River. In *Encyclopedia of Deserts* (ed. M.A. Mares) p. 129. University of Oklahoma Press, Norman.

Chatty, D. (1996). *Mobile pastoralists: development, planning and social change in Oman*. Columbia University Press, New York

Cleland, J.H., York, A., and Johnson, A. (2000). The tides of history: modeling Native American use of recessional shorelines. In *ESRI 2000 User Conference Proceedings*, ESRI, Redlands, California

Clotts, H.V. (1917). *History of the Papago Indians and history of irrigation, Papago Indian Reservation, Arizona*. Department of the Interior, United States Indian Service, Washington, D.C.

De Villiers, M. (2000). *Water*. Stoddart Publishing Co. Limited, Toronto.

Diamond, J. (2005). *Collapse: how societies choose to fail or succeed*. Viking, New York.

Dregne, H.E. (1999a). Nile River. In *Encyclopedia of Deserts* (ed. M.A. Mares) pp. 391–392. University of Oklahoma Press, Norman.

Dregne, H.E. (1999b). Tigris-Euphrates. In *Encyclopedia of Deserts* (ed. M.A. Mares) p. 561. University of Oklahoma Press, Norman.

Ebraheem, A.M., Riad, S., Wycisk, P., and Sefelnasr, A.M. (2004). A local-scale groundwater flow model for groundwater resources management in Dakhla Oasis, SW Egypt. *Hydrogeology Journal* 12(6): 714–722

Ezcurra, E., and Rodrígues, V. (1986). Rainfall patterns in the Gran Desierto, Sonora, Mexico. *Journal of Arid Environments* 10:13-28.

Felger, R.S. (2006). Living resources at the center of the Sonoran Desert: Native American plant and animal utilization. In *Dry Borders: Great Natural Reserves of the Sonoran Desert*. (eds. R.S. Felger, B. Broyles). University of Utah Press, Salt Lake City.

Felger, R.S., and Moser, M.B. (1985). *People of the Desert and Sea: Ethnobotany of the Seri Indians*. University of Arizona Press, Tucson.

Felger, R.S., and Nabhan, G.P. (1978). Agroecosystem diversity: A model from the Sonoran Desert. In *Social and Technological Management in Dry Lands. AAAS Selected Symposium 10*. (ed. N.L. Gonzalez) p. 128-149. Westview Press, Boulder, Colorado.

Fish, S. (2004). Corn, crops, and cultivation in the Greater Southwest. In *People and Plants of Ancient Western North America* (ed. P. Minnis) p. 115-166. Smithsonian Institution Press, Washington, D.C.

Fish, S., Fish, and P.R., Marsden, J. (1992). Evidence for large scale agave cultivation in the Marana community. *In The Marana Community in the Hohokam world*. (eds. S. Fish, P. Fish, J. Marsden) p. 73-87. University of Arizona Press, Tucson.

Flegg, F. (1993). *Desert: A Miracle of Life*. Blandford, London.

Flint, R., and Flint, S. (editors and translators). (2005). *Documents of the Coronado Expedition, 1539-1542*. Southern Methodist University Press, Dallas.

Ford, R. I. (1983). Inter-Indian exchange in the Southwest. In *Southwest (ed. A. Ortiz) p. 711-722. Handbook of North American Indians,* vol. 10. Smithsonian Institution, Washington, D.C.

Gentry, H.S. (1982). *The Agaves of Continental North America*. University of Arizona Press, Tucson.

Golany, G. (1979). 'Israeli Development Policies and Strategies in Arid Zone Planning'. In *Arid Zone Settlement Planning: The Israeli Experience* (ed. G. Golany) pp. 3–42. Pergamon Press, New York

Gordon, R.J., and Douglas, S.S. (2000). *The Bushman Myth: The Making of a Namibian Underclass*. Westview Press, Oxford.

Halpern, E.A. (1999). Camel. In *Encyclopedia of Deserts* (ed. M.A. Mares) pp. 95–97. University of Oklahoma Press, Norman.

Haury, E.W. (1975). Shells. In *Excavations at Snaketown* (eds. H.S. Gladwin, E.W. Haury, E.B. Styles, N. Gladwin) p. 135-153. University of Arizona Press, Tucson. Reprint of 1938, Medallion Papers XXV, Gila Pueblo, Globe, Arizona.

Henschel, J.R., Burke, A., and Seely, M. (2005). Temporal and spatial variability of grass productivity in the central Namib Desert. *African Study Monographs* Suppl. 30: 43–56.

Henschel, J.R., Dausab, R., Moser, P., and Pallett, J. (eds). (2004). *!nara: fruit for development of the !Khuiseb Topnaar*. Namibia Scientific Society, Windhoek.

Hodgson, W.C. (2001). *Food Plants of the Sonoran Desert*. University of Arizona Press, Tucson.

Hulett, G.K., and Charles, A.R. (1999a). Death Valley. In *Encyclopedia of Deserts* (ed. M.A. Mares) pp. 147–148. University of Oklahoma Press, Norman.

Hulett, G.K., and Charles, A.R. (1999b). International Biological Programme. In *Encyclopedia of Deserts* (ed. M.A. Mares) pp. 301–302. University of Oklahoma Press, Norman.

Issar, A. (1999). The Past as a Key for the Future in Resettling the Desert. In *Desert Regions: Population, Migration, and Environment* (eds. B.A. Portnov and A.P. Hare) pp. 241–248. Springer Verlag, Heidelberg

Jacobsohn, M. (1994). Southern Africa – people of the Namib. In *Deserts* (ed. M. Seely) pp. 100–102. The Illustrated Library of the Earth, Weldon Owen, Sydney.

Kates, R.W., Johnson, D.L., and Johnson-Haring, K. (1977). Population, Society and Desertification. In *Desertification: Its Causes and Consequences* (ed. United Nations Conference on Desertification) pp. 261–318. Pergamon Press, Oxford

Kimber, R.G. (2005). 'Because it is our country': The Pintupi and their return to their country, 1970–1990. pp. 345–356. In *23°S Archaeology and Environmental History of the Southern Deserts*. National Museum of Australia Press, Canberra.

Kinahan, Jill (2000). Cattle for beads. *Studies in African Archaeology 17*. Uppsala University, Sweden and Namibia Archaeological Trust, Windhoek.

Kinahan, John (1991). *Pastoral Nomads of the Central Namib Desert: The People History Forgot*. Namibia Archaeological Trust & New Namibia Books, Windhoek.

Kinahan, John (2005). The late Holocene human ecology of the Namib Desert. In *23°S Archaeology and Environmental History of the Southern Deserts*. (ed. M. Smith and P. Hesse) pp. 120–131. National Museum of Australia Press, Canberra.

Lacher, T.E. Jr. (1999). Off-road vehicles. In *Encyclopedia of Deserts* (ed. M.A. Mares) pp. 396–397. University of Oklahoma Press, Norman.

Lancaster, W.O. (1981). *The Rwala Bedouin today*. Cambridge University Press, Cambridge

Louw, G., and Seely, M. (1982). *Ecology of Desert Organisms*. Longman, New York.

Lövenstein, H.M. (1994). Agricultural development. In *Deserts* (ed. M. Seely) pp. 151–155. The Illustrated Library of the Earth, Weldon Owen, Sydney.

Luedeling, E., Nagieb, A., Wichern, F., Brandt, M., Deurer, M., and Buerkert, A. (2005). Drainage, salt leaching and physico-chemical properties of irrigated man-made terrace soils in a mountain oasis of northern Oman. *Geoderma* 125(3–4): 273–285

Mares, E.A. (1999). Religion in deserts. In *Encyclopedia of Deserts* (ed. M.A. Mares) pp. 471–472. University of Oklahoma Press, Norman.

Marlett, C.M. (2005), Personal communication

Marx, E. (1977). Communal and individual pilgrimage: the region of saints' tombs in South Sinai. In *Regional Cults* (ed. R. Webner) pp. 29-54, Academic Press, Londonß

Marx, E. (1994). The Middle East. In *Deserts* (ed. M. Seely) pp. 110 –115. The Illustrated Library of the Earth, Weldon Owen, Sydney.

McGuire, R.H., and Schiffer, M.B. (editors). (1982). *Hohokam and Patayan: Prehistory of Southwestern Arizona*. Academic Press, New York.

Montaña, E, Torres, L., Abraham, E.M., Torres, E., and Pastor, G. (2005). Los espacios invisibles. Subordinación, marginalidad y exclusión de los territorios no irrigados en las tierras secas de Mendoza, Argentina. *Región y Sociedad* (El Colegio de Sonora, México) 17(32): 4–32

Moore, K.M., Miller, N.F., Hiebert, F.T., and Meadow, R.H. (1994). Agriculture and herding in the early oasis settlements of the Oxus civilization. *Antiquity* 68(259): 418–427

Nabhan, G.P. (1985). *Gathering the Desert*. University of Arizona Press, Tucson.

NACSO. (2004). *Namibia's Communal Conservancies: A Review of Progress and Challenges*. NACSO, Windhoek.

Ngaido, T., Shomo, F. and Arab, G. (2001). Institutional change in the Syrian rangelands. *IDS Bulletin* (Institute of Development Studies) 32(4): 64–70

Ocampo, O. (1999). Movies in deserts. In *Encyclopedia of Deserts* (ed. M.A. Mares) pp. 378–382. University of Oklahoma Press, Norman.

Pailes, R.A. (1999). Desert peoples. In *Encyclopedia of Deserts* (ed. M.A. Mares) pp. 156–167. University of Oklahoma Press, Norman.

Parrish, A.M. (1995). 'There were no sus in the old days:' post-harvest pest management in an Egyptian oasis village. *Human Organization* 54(2): 195–204

Portnov B.A., and Safriel, U. (2004). Combating desertification in the Negev: dryland agriculture vs. dryland urbanization. *Journal of Arid Environment* 56(4): 659–680

Portnov, B.A., and Erell, E. (1998). Long-term Development Peculiarities of Peripheral Desert Settlements: The Case of Israel. In *International Journal of Urban and Regional Research* 22(2): 216–32

Rea, A.M. (1981). Resource utilization and food taboos of Sonoran Desert peoples. *Journal of Ethnobiology* 1:69-83.

Rea, A.M. (1983). *Once a River: Bird Life and Habitat Changes on the Middle Gila*. Tucson: University of Arizona Press.

Rea, A.M. (1997). *At the Desert's Green Edge: An Ethnobotany of the Gila River Pima*. University of Arizona Press, Tucson.

Reader, J. (1997). *Africa: A Biography of the Continent*. Penguin Books, London.

Reisner, M. (1986). *The Cadillac Desert: The American West and its Disappearing Water*. Penguin, New York.

Roig, F.A., Gonzalez-Loyarte, M., Abraham, E.M., Méndez, E., Roig, V.G., and Martínez Carretero, E. (1991). Maps of desertification hazards of Central Western Argentina (Mendoza Province). In *World Atlas of Thematic Indicators of Desertification* (ed. UNEP). Edward Arnold, London

Saini, B. S. (1980). *Building in Hot Dry Climates,* John Wiley & Sons, Chichester

Sandelowsky, B.H. (1974). Archaeological investigations at Mirabib Hill rock shelter. *The South African Archaeological Society Goodwin Series* 2: 65–72.

Seely, M. (2004). *The Namib: Natural History of an Ancient Desert.* DRFN, Windhoek.

Shackley, M. (1980). An Acheulean industry with *Elephas recki* fauna from Namib IV, South West Africa (Namibia). *Nature* 284 (5754): 340–341.

Shackley M. (1983). Human burials in hut circles at Sylvia Hill, S.W. Africa/Namibia. *Cimbebasia Series B* 3 (3): 102–106.

Shreve, F. (1951). *Vegetation of the Sonoran Desert.* Carnegie Institution of Washington Publication No. 591. Reprinted 1964, part 1 of F. Shreve, I. L. Wiggins, *Flora and Vegetation of the Sonoran Desert.* Stanford University Press, Stanford.

Smith, A.B. (1994) African desert people: North Africa. In *Deserts* (ed. M. Seely) pp. 94–98. The Illustrated Library of the Earth, Weldon Owen, Sydney.

Smith, M., and Hesse, P. (eds) (2005). *23°S Archaeology and Environmental History of the Southern Deserts.* National Museum of Australia Press, Canberra.

van den Eynden, V., Vernemmen, P., and van Damme, P. (1992). *The Ethnobotany of the Topnaar*. Universiteit Gent.

Wang, B.C., Qiu, H.X., Xu, Q., Zheng, X.L., and Liu, G.Q. (2000). The mechanism of groundwater salinization and its control in the Yaoba Oasis, Inner Mongolia. *Acta Geologica Sinica*-English Edition 74(2): 362–369

Wehner, R. (1990). On the brink of introducing sensory ecology: Felix Santschi (1872–1940) – Tabib-en-Neml. *Behavioural Ecology and Sociobiology* 27:295–306.

Wrigley, G. (1995). Date palm *Phoenix dactylifera*. In *Evolution of crop plants,* 2nd ed. (eds. J. Smartt and N. W. Simmonds) pp. 399-403. Longman, Harlow, UK

Zohary, D. and Hopf, M. (2000). *Domestication of plants in the Old World* 3rd ed. Oxford University Press, Oxford, UK

3

Chapter 3: Deserts and the Planet — Linkages between Deserts and Non-Deserts

Lead author: Uriel Safriel
Contributing authors: Exequiel Ezcurra, Ina Tegen, William H. Schlesinger, Christian Nellemann, Niels H. Batjes, David Dent, Elli Groner, Scott Morrison, Danny Rosenfeld, Uzi Avner, Noah Brosch, Avi Golan-Goldhirsh, Pinchas Alpert, Boris A. Portnov, Rex Cates, Robin P. White, Anastasios Tsonis, Moshe Schwartz, Yoram Ayal, Berry Pinshow, Dan Cohen, Thomas Deméré, Haim Shafir, Andrew Warren, Emanuel Mazor

This chapter addresses the linkages and interactions between deserts and the rest of our planet. While desert climate is controlled by processes taking place outside deserts, processes in deserts also affect climate away from deserts. Deserts and non-deserts are linked by dust generated in deserts that travels away from deserts, and rivers that originate outside deserts dramatically affect deserts while flowing through them. People from outside deserts migrate or visit deserts, while desert people export minerals and fossil energy to non-desert economies. Deserts also serve as corridors through which goods travel and cultures are exchanged, and desert corridors serve bird migration and locust invasions. Finally, though most deserts are remote from leading centres of science, research carried out in deserts has enriched knowledge of the history of our universe and planet, of life on earth, and of peoples and their cultures.

The Physical Tele-Connections — Climate, Dust and Rivers

Deserts are not only highly constrained by water, but their water supply greatly depends on climatic processes operating away from them, either those involved in the generation of desert rainfall, or in generating flows of rivers that enter deserts. Deserts also have effects on climate beyond their boundaries, either directly or through the dust they generate.

DESERTS AND GLOBAL CLIMATE

Ocean-atmosphere linkages maintain desert climate

The intense solar radiation hitting the earth of the tropical belt (between latitudes 23° South and 23° North) sets off air currents bringing dry winds to sub-tropical areas (within latitude 25° and latitude 35°, either North or South), thus denying them precipitation and making them deserts. The western seaboard sections of the African and American continents are deserts due to dry inland conditions induced by upwelling of deep cold seawater, driven by ocean coastal currents. Deserts also occur on the leeward sides of mountain ranges that are deprived of ocean-generated moisture. It is rather paradoxical that the

forces which induce highly productive conditions next to deserts — the high solar radiation in the tropical rainforests, the cold and nutrient-rich upwelling of western coastal seawaters, and the moisture laden tropical trade winds reaching tropical continental mountains — are also all factors maintaining the aridity of deserts.

Rainfall patterns within deserts also depend on climatic processes outside deserts. When the desert surface is cold, moisture blown from the sea condenses and generates winter rains; when the surface is warm, the moisture drawn into the atmosphere from various sources condenses to generate summer rains. Rainfall can be also augmented by of fog, formed when water droplets kept in suspension over the ocean are blown into the desert.

The great year-to-year variations in desert rainfall are modulated by processes away from deserts, such as the Southern Oscillation — a global weather cycle associated with a fluctuation of atmospheric pressure between the South Pacific and tropical Indian Oceans, and expressed in alternating 3–7 year cycles of El Niño Southern Oscillation (or ENSO) and La Niña events. Within this cycle, El Niño develops when the warm water that accumulates in the eastern equatorial Pacific Ocean decreases the upwelling of cold water along the coasts of North and South America; the sea surface then warms-up, resulting in an increase in coastal winter rainfall. After a few months, El Niño conditions begin to recede, paving the way for La Niña: a strong westward flow of surface currents forces the upwelling of cold waters along the coasts of the American continent, thus inducing drought along the coastal deserts. This cycle affects the coastal deserts of Atacama and Baja California, Namibia, Western Australia, and Atlantic Morocco. Paradoxically, since deep seawater upwelling also modulates the productivity of coastal water, the pulse of rainfall-induced high desert productivity coincides with a pulse of low coastal ocean productivity, and vice versa (see also Chapter 1).

Because the El Niño-induced coastal rainfall occurs when moist air moves from warm oceans onto cooler land, the increase in precipitation occurs mostly in coastal fog deserts such as the southern Namib or the Atacama deserts. Inland, summer-rain deserts such as central Australia, the Thar Desert in India and Pakistan, or the Brazilian Caatinga, however, tend to become drier during El Niño years but enjoy increased precipitation during La Niña years, due to the differential in temperatures between these desert lands and the surrounding seas. Because deserts are so limited by scanty rainfall, these year-to-year climatic oscillations produce pulses of abundance and scarcity of resources with immense ecological repercussions. This highlights the importance of tele-connections in global ecology: a gradient of air pressures developing over the sea critically affects the ecological dynamics of the driest of lands.

Global climate change affects desert climate

Global climate change, the directional change induced by anthropogenic emissions of greenhouse gases (to be distinguished from long-term or short-term climate variations not caused by global-scale human impact on the climate system) also affects deserts. Deserts warmed-up between 1976 to 2000 at an average rate of 0.2–0.8°C/decade — an overall increase of 0.5–2°C (Table 3.1), much higher than the average global temperature increase of 0.45°C, which has been attributed to the increase in atmospheric concentrations of greenhouse gases (IPCC 2001). Global warming is expected to induce an overall increase in rainfall; but high latitudes are projected to warm more than the mid- and low-latitudes, resulting in more rainfall in higher latitudes linked to reduced rainfall in subtropical ones. Indeed, in most deserts within the subtropical belt, rainfall has already been decreasing in the last two decades (Table 3.1).

Table 3.1: Changes in desert temperatures and rainfall

Observed and projected changes in surface temperatures (ΔT, °C) and annual precipitation (Δ rainfall, per cent) over twelve desert regions. A2 and B2 are IPCC SRES scenarios (IPCC 2001). Empty cells indicate no data. Values in bold highlight deserts for which no decrease and/or an increase in precipitation is predicted. Values given for observed changes per decade are the average of two values: (i) the difference between 1981-1990 mean values and 1976-1980 mean values, and (ii) the difference between 1991-2000 mean values and 1981-1990 mean values (percentage diffferences for rainfall).

Desert (country)	Observed changes 1976-2000		Projected changes (means of values projected for 2071-2100 relative to means of observed 1961-1990 data)			
	ΔT (°C/decade)	Δ rainfall (%/decade)	ΔT (°C)		Δ rainfall (%)	
			A2 scenario	B2 scenario	A2 scenario	B2 scenario
Sahara (Algeria, Tunisia, Morocco)	–	–	+4	+3	-20	-10
Libya (Libya)	+0.8	–	+3	+2	-10	-10
Western Desert (Egypt)	+0.8	+4	+4	+2	**0**	**0**
Rub' al Khali (Saudi Arabia)	–	–	+4	+3	**+2**	**+2**
Kalahari (South Africa)	+0.8	-12	+4	+3	**+10**	**0**
Namib (Namibia)	-0.4	–	+3	+2	**0**	-5
Gobi (China)	+0.8	–	+6	+4	**+15**	**+10**
Kizil Kum (Afghanistan)	+0.6	+8	+7	+5	-5	**+5**
Dashti Kbir (Iran)	+0.6	-16	+5	+4	-10	**+10**
Great Victoria (Australia)	+0.8	–	+3	+2	-10	-5
Colorado and Great Basin (USA)	+0.4	-4	+5	+3	-15	-5
Atacama (Chile)	+0.2	-8	+2	+1	-10	-5

Source: IPCC 2001

average rate of accumulation of the inorganic carbon is considerably slower — some 0.1–0.6 g C m^{-2} y^{-1} (Schlesinger 1985). Yet, though the rate at which atmospheric CO_2 is precipitated as soil carbonate is slow, and the rate at which it can be sequestered as soil organic matter is relatively fast, the global desert ecosystem has accumulated much inorganic carbon but only little organic carbon. Thus, only 9–10 per cent of global soil organic carbon is held in deserts. Therefore, in spite of their large extent, deserts do not play a significant role in the global carbon cycle.

However, if some desert regions do become significantly moister under global warming, they have the potential to function as a globally significant sink that could tangibly mitigate global warming (Lioubimtseva and Adams 2004). On the other hand, those deserts that become drier, with their vegetation only weakly responding to CO_2 enrichment, will not become a significant sink. These deserts are also not likely to act as a significant source driven by land degradation, because the turnover rate of the large desert sink of inorganic soil carbon is too slow to generate significant CO_2 emissions. Also, although the turnover of soil organic carbon is fast and land degradation in deserts might increase CO_2 emissions (as the carbon in eroded soil is oxidized), the pool of soil organic carbon that might be affected by land degradation is too small to make this a significant contribution to global atmospheric CO_2. Between-ecosystem comparison (Batjes and Sombroek 1997) suggests that under the scenario of further desert warming and reduced precipitation, the ratio of soil organic carbon to soil inorganic carbon in deserts will be reduced. This will further reduce the role of deserts in the modulation of global climate change, unless drastic human interventions for increasing desert productivity take place.

DESERTS AND DUST

How, where and when is desert dust formed and where does it travel to?

Deserts generate dust, much of which travels great distances into non-desert areas, with diverse and often unexpected effects. Far-travelled dust particles are usually less than 2 micrometres (µm) in size, and are mostly made up of an alumino-silicate minerals. The major desert dust production mechanism is "saltation", a process triggered when larger wind-blown particles bounce on the desert soil's surface, thus releasing smaller dust particles from the surface. Dust is emitted from the Sahara, Arabian, Gobi, Taklimakan, Australian and South American deserts; but, quantitatively, most dust in the global atmosphere is emitted from the hyperarid northern African (50–70%) and Asian (10–25%) deserts.

Frequent dust events are observed in enclosed depressions (Prospero and others 2002): from lake sediments deposited during wetter climate periods (like the Paleo-Lake Chad on the Saharan–Sahelian border, which contains the most active dust source on earth) or from the end-points of riverine transport of fine particles (like the Murray-Darling Basin in Australia). Global annual dust emissions are estimated to range from 1 000 to 3 000 million tonnes per year (IPCC 2001), less than 10 per cent of which is likely to result from human activities in the drylands (Tegen and others 2004).

Dust can be carried over thousands of kilometres by strong winds (Figures 3.1 and 3.2). Dust emitted in the Sahara can be carried across the North Atlantic to North and Central America, and even to the Amazon basin. Large amounts of Asian dust are carried over the North Pacific toward the mid-Pacific islands and North America. The lifetimes of atmospheric dust range from less than a few hours for particles larger than 10 µm, which are quickly removed by gravitational settling, to 10–15 days for submicron particles that are mostly removed by wet deposition (Jickells and others 2005).

Desert-generated dust affects productivity of land and ocean away from deserts

Dust generated in deserts adds essential nutrients to terrestrial and marine ecosystems away from deserts, such as phosphorus and silicon, which enhance growth in oceanic phytoplankton otherwise often limited by these minerals. Iron is a micronutrient whose shortage limits the uptake and assimilation of nitrogen, phosphorus and silicon. Enrichment by dust-carried iron can stimulate oceanic plankton growth, and therefore increase CO_2 uptake in ocean regions, where iron is limiting. In nutrient-poor regions, dust-borne iron

Figure 3.1: Atmospheric dust

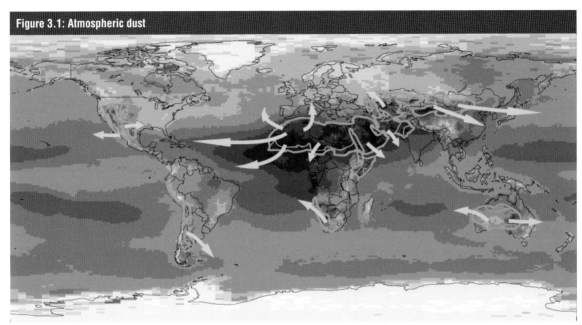

Twelve-year annual average of the Absorbing Aerosol Index (AAI) from the Total Ozone Mapping Spectrometer (TOMS) satellite instrument. The shading provides a qualitative measure of the atmospheric dust content: dark shade – high aerosol concentration. The TOMS AAI is an indicator of the load of absorbing aerosols in the atmosphere, which are mostly dust particles with some contribution of smoke particles. The desert areas are marked by contours, main transport pathways of dust out of the deserts are marked by arrows.

Source: NASA's "Blue Marble" website

Figure 3.2: Travelling dust storms

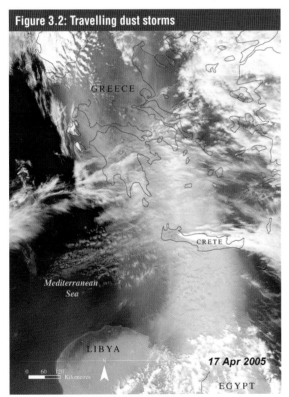

Dust storm moving over the Mediterranean from the northern Sahara to southern Europe.

Source: NASA's "Blue Marble" website, http://earthobservatory.nasa.gov

can enhance the fixation of molecular nitrogen by phytoplankton. However, because the specific iron compounds required by phytoplankton are at low concentration in dust and are not easily soluble in water, the role of dust-borne iron in ocean productivity is not yet clear (Jickells and others 2005). Transported dust may also have negative oceanic effects: some authors argue that increased dust deposition in the western Atlantic over the past 25 years could have significantly contributed to coral reef decline by carrying bacterial or fungal spores (Shinn and others 2000). On the other hand, phosphate deposited by dust enhances forests of the south-eastern United States, and the Saharan dust deposited in the Amazon basin replenishes the phosphorus lost through the intense leaching caused by high rainfalls in this area (Okin and others 2004).

Desert dust affects atmospheric properties, rainfall, visibility, and health away from deserts

Depending on their size, distribution and refractive properties, dust particles in the atmosphere partly reflect and partly absorb incoming solar radiation (Sokolik and others 2001). Thus, dust blown away from deserts and over oceans increases the reflectance in an area in which the dark ocean surface would otherwise be absorbing radiation, and thus the atmosphere over the oceans is cooled. When desert dust reaches heights above 5 km, it absorbs and reflects back to space some of the solar radiation, and so warms the mid-troposphere (Kishcha and others 2003) at the expense of cooling the lowest levels. This generates a downward airflow that exacerbates desert conditions. The added dryness can lead to more desert dust, thus amplifying the initial effect. Desert dust particles can impair precipitation from potential rain clouds, and keep the desert drier, dustier and even less favourable to precipitation in a reinforcing feedback loop, which further increases dust generation by deserts and the likelihood of its transport to non-deserts. Far away from deserts, the transported dust may suppress precipitation from convective clouds by inhibiting the formation of raindrops (Rosenfeld and others 2001). Finally, desert-generated dust may reduce visibility to the point of seriously interfering with ground and air traffic away from deserts. Persistent dust storms also increase the incidence of respiratory diseases (Gyan and others 2005).

Desert dust and global climate change

In general, both climate change-induced increasing aridity of deserts and increasing wind speeds will increase overall dust emissions from deserts. In deserts where rainfall is predicted to decrease, concurrent loss of vegetation cover will allow more dust emissions to non-desert areas. In deserts where rainfall is predicted to increase, desert dust flux will be reduced, sustaining, in turn, wet conditions away from deserts (Lioubimtseva and Adams 2004). Yet, due to uncertainties, projections of dust emissions for the next 100 years range between a 60 per cent decrease to a 50 per cent increase in dust emissions (Mahowald and Luo 2003).

CROSS-DESERT RIVERS OF NON-DESERT SOURCES
The significance of rivers to the deserts they cross

Low and variable precipitation and high evaporation are not conducive to generating perennial rivers in deserts. Rather, the source of perennial rivers in deserts is from upland, non-desert areas. The headwaters of the Nile which crosses the eastern Sahara Desert, the Gariep river of the Kalahari, the Tigris and Euphrates in the Syrian Desert, the Indus of the Thar Desert, and other desert rivers, are far from the desert edge, up in humid highlands (Figure 3.3). At a certain point of their course these rivers cross a desert to eventually discharge into the sea (like the Colorado and the Tigris rivers), or into a desert lake (like the Amu Darya and Syr Darya rivers). While flowing through the desert, the riparian banks function as elongated, winding oases, a non-desert island within the desert. Cross-desert rivers provide fish, plants, and animals that comprise the base of

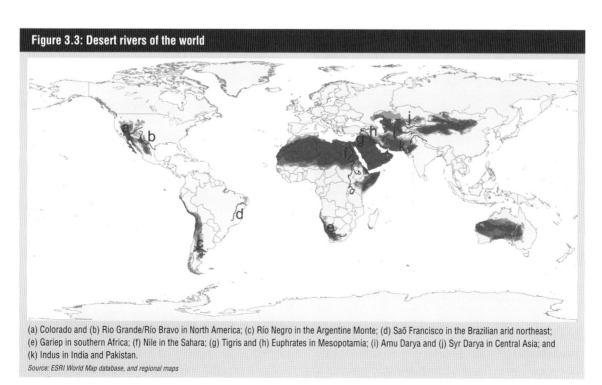

Figure 3.3: Desert rivers of the world

(a) Colorado and (b) Rio Grande/Río Bravo in North America; (c) Río Negro in the Argentine Monte; (d) São Francisco in the Brazilian arid northeast; (e) Gariep in southern Africa; (f) Nile in the Sahara; (g) Tigris and (h) Euphrates in Mesopotamia; (i) Amu Darya and (j) Syr Darya in Central Asia; and (k) Indus in India and Pakistan.
Source: ESRI World Map database, and regional maps

livelihoods for people concentrating along their courses (for example, 97 per cent of the Egyptian population lives along the Nile). River water is often diverted to irrigate extensive agriculture and to support pasture, and it carries sediment that fertilizes or generates soil. Desert rivers have cultural and spiritual significance for desert people, nurturing ancient civilizations.

Climate away from deserts modulates the flow of desert rivers

The flow of perennial rivers in deserts totally depends upon the upland headwaters, non-desert wetlands and lakes, and on their pre-desert course. For example, the Egyptian population, most of which lives under climatically hyperarid conditions with far less than 100 mm of annual rainfall, totally depends on a rainfall regime of more than 1 600 mm that precipitates some 3 000–4 000 km away. The Mesopotamian rivers cross a desert of less than 200 mm of annual rainfall, but depend on precipitation of more than 1 000 mm, much of which is snowfall maintaining seasonal peaks of desert flow, depending on the timing and rate of snowmelt 900 km away. Desert flows of rivers such as the Nile and the Colorado are extremely sensitive to variations in rainfall interception within their small headwater catchments outside the desert (Degens and others 1990).

Water diversion away from deserts reduces cross-desert flow

The amount and quality of river water reaching deserts often depend more on people than on nature. The river flows in the Egyptian and Iraqi deserts depend on the management of their headwaters in Ethiopia and Turkey, respectively. In countries like Pakistan, with both desert and non-desert areas, desert people depend on the way water is managed in the non-desert sections of the Indus river, which has a denser population and a greater impact on the river-flow. Damming that diverts water for irrigation and generation of electric power in the upland non-desert courses of rivers reduces the amount of water reaching the desert. Conflicts between highland and lowland water users are becoming common globally, but they are more apparent in desert-crossing rivers, because most deserts are in the lower river reaches, where

water is more precious and population growth is often high; most dams in the Tigris-Euphrates basin are in the Turkish, non-desert course of the two rivers, while few lie in the desert Syrian and Iraqi sections, where flow has been much reduced. Yet, such conflicts have not escalated into armed confrontations; rather they often motivate cooperation among the riparian countries.

Water quality of cross-desert rivers depends on humans away from deserts

Deforestation and overgrazing at the source enriches river water with minerals leached or transported with eroded soils (Hanspeter and others 1998). Residual pesticides and fertilizers, irrigation-generated salinity, and industrial and organic wastes are also drained into river flows. Much of the organic pollution dumped in the non-desert section oxidizes before it reaches the desert; but other pollutants do reach the desert flow, which lacks incoming tributaries to dilute pollutants or spring floods to wash them away. Since the damming of most desert rivers, chiefly off-desert, reduces sediment load in the river flow and hence nutrients, fisheries and wildlife have been impacted. For example, the demise of the Mesopotamian Marshlands is in part due to such reduced flows (Richardson and others, 2005).

Effects of global climate change in the non-desert source areas

Global climate change will affect the remote sources of desert rivers more than deserts themselves. Those rivers with headwaters in snow-capped mountains that depend almost exclusively on snowmelt from glaciers — such as on both sides of the Andean range (Atacama and Monte) — will be the most affected, because accelerated melting of most glaciers is predicted with high confidence (IPCC 2001). The Himalayan glaciers, surrounded by relatively dry areas and sustained due to the high elevations where water is stored as ice, are highly vulnerable; melt will first generate increased flow and the eventual loss of the glaciers would reduce desert flow dramatically. With most of Pakistan's inhabitants dependent upon an irrigation network now fed by the Indus river, the effects of climate change in its basin could be devastating (Nianthi and Husain 2004). The Nile catchment is located at the boundary

housing developers, etc. — which create additional employment opportunities for desert newcomers.

Migration to the desert may also take place for security considerations. Thus, during WWII, major industries were relocated from the western part of the former Soviet Union, occupied by Nazi Germany, to its eastern regions, including the deserts of the Kazakh and Turkmen republics. This relocation was followed by a major migration of technical personnel and employees of these industries. In recent years, the government of China established incentives promoting primary and military industries to boost the economy of its western and northern desert regions, driven by the discovery of oil in these regions and by a policy of encouraging development of the inland parts of the country. Other policy-encouraged immigration into deserts, which may affect smaller desert societies, their lifestyles, cultures and environment, are the immigration of Han Chinese into the Uyghur-inhabited Xinjiang (Nellemann 2005), and the immigration of Delta-inhabiting Egyptians into the small Bedouin societies of the Sinai, as well as the settling of Egyptian university graduates in remote desert localities charged with reclaiming them for cultivation (Divon and Abou-Hadab 1996).

In industrial countries, migration from non-desert to desert areas is driven by the availability of cheap housing (development towns in the Negev Desert of Israel), including for retired citizens (the Sun Belt localities in the US, or the Canary Islands) who are attracted to desert towns by the dry and sunny climate. In developing desert countries, specifically in Sub-Saharan Africa, periodic droughts in non-desert drylands draw thousands of rural migrants and nomads to local cities, many of which are located adjacent to deserts, in search of food and employment (Pedersen 1995) (see also Chapter 2).

Tourist influxes to deserts encourage migration to deserts

In recent years, many desert areas south of the Mediterranean basin (e.g., Canary Islands, Eilat in Israel, Sharm-al-Sheikh in Egypt), have become popular destinations for tourists from northern countries, who are attracted by the balmy climate of the desert. Many desert resorts in the Mediterranean spur their attractiveness by combining recreational

facilities for vacationers with visits to adjacent archaeological and geological parks. Rehabilitation centres for patients suffering from diseases, such as asthma or arthritis, have also been established in desert regions (Golany 1978). Desert tourism boosts the economy of desert countries (11 per cent of Egypt's gross national income is from tourism; WRI 2003), and services for the desert tourism industry also create new jobs and attract immigration into the growing desert cities.

Deserts as Corridors

The common denominator of trading caravans and migratory birds is that both use deserts only as corridors linking non-desert starting points with non-desert destinations. However, while using the desert corridor, both often interact with the desert environment, its biota and its human population. And, just as desert corridors have been used by armies to reach their non-desert destinations for assault, so it is often the case with locusts that use deserts as a corridor for invading non-desert areas.

CROSS-DESERT TRANSPORTATION OF GOODS AND CULTURES

Deserts are wedged between, and thus hinder exchanges among, civilizations. In response, desert people developed a livelihood that capitalized on this need for commercial exchanges, the livelihood of transportation — guiding and servicing the cross-desert caravans. This expertise in safely and efficiently moving goods through deserts channelled a flow of income from non-desert to desert people, and at the same time economically and culturally benefited the non-desert areas at both ends of the cross-desert transportation routes. This trade often made desert people knowledgeable of the politics of Europe and Asia, more than the other way around. Though the great trade empires founded on cross-desert transport are long gone and desert routes are now far less significant, transport and trade still support desert livelihoods.

Deserts have been crossed by trade routes through millennia

Most deserts have been crossed by trading roads through millennia (Figure 3.4). The Silk Roads were already active in the late Bronze Age, though

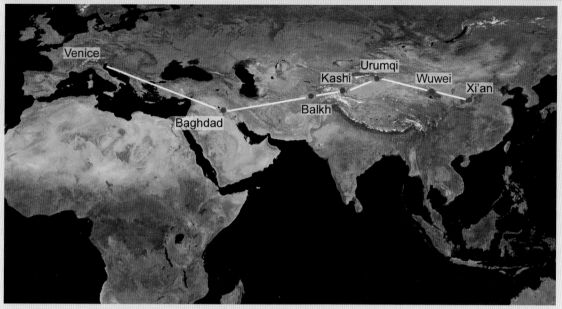

Figure 3.4: The Silk Road: trans-desert trade route in Eurasia, 3rd–15th century

Source: Modified from Donnus Nicolaus Germanus, Cosmographia: Claudius Ptolemaeus; Leonhart Holle, Ulm, 1482

intensive use of cross-desert roads was triggered by the domestication of the camel. The trade through the trans-Saharan roads took off only with the Islamic conversion of West Africa. The two main roads, made of a network of shorter segments between oases, led from Morocco to the Niger Bend and from Tunisia to Lake Chad. Guided by Berber guides to ensure safe passage, caravans included on the average a thousand camels, sometimes reaching 12 000 animals, and runners were sent ahead to oases to ship out water when the caravan was still days away. West African gold and slave servants were exchanged for North African salt and slave soldiers, thus enriching kingdoms and empires of Ghana and Mali south of the Sahara, and Tuareg cities north of it. Similarly, through the Silk Road network goods to and from Xinjiang province of China travelled through Central Asian deserts either to West Asia or to Russia.

Trade through these roads declined (as of the 16th and the 12th centuries in Africa and Asia, respectively) due to political unrest, incursions, and wars, on the one hand, and the development of maritime routes on the other. The independence of African nations in the 1960s and rebellions and civil wars of the 1990s halted the cross-Saharan roads at the national boundaries, and trade through the Silk Road was disrupted by the

wars of Genghis Khan. Today most cross-desert transport is through an extensive tarmac road network in addition to transport by air and sea; yet, Tuareg camel caravans still travel on the traditional Saharan routes, carrying salt from the desert interior to communities on the desert edges.

Cross-desert trade routes encouraged significant cultural exchange

The transfer of goods between non-desert lands through deserts enriched desert people, both economically and culturally. The Nabatean Kingdom was moulded and subsisted on controlling roads crossing West Asian deserts, moving spices from southern Arabia and goods from India to their capital Petra, and then to the Mediterranean port of Gaza, to be shipped to Greece and Rome. Other desert trading cultures include those of the Saharan Tuaregs, Fulani and Songhai and the Central Asian Uyghurs and Kazaks.

The cross-desert trading routes functioned as communication and information channels between non-desert regions, and between these and the desert people. Through the Sahara rumours of West African treasures prompted the Portuguese to reach Guinea, and West Africans became acquainted with the Arab and Mediterranean world long before the adoption of Islam. Books from Europe travelled to Africa through the trans-Saharan roads, the only means for their transport until the 15th Century, exposing sub-Saharan countries to knowledge generated in Europe (Masonen 1997). European traders met trading partners in Africa and Asia with information far surpassing their own. Along with luxury goods and weapons, religions and knowledge moved through and out of the Asian deserts, such that the cultural scene of Central Asia was moulded by Indian, Greek, Chinese, Tibetan and Arabian cultures and the Buddhist, Christian and Muslim religions. Cross-desert trade routes also promoted gene flow among populations isolated from each other by the desert's reproductive barrier. Shared gene pools among camels, horses and goats on both sides of deserts (Jianlin and others 2004) are attributed to Pleistocene migrations along corridors later to become the Silk Road. Human populations in Europe, Northern Africa, the Arabian Peninsula, East Asia and North America share a common genetic history, attributed to cross-desert travel and trading (Yao and others 2004).

Cross-desert transport continues to affect desert and non-desert people

While the trade across the desert today is small, cross-desert pipelines, trains and trucks move minerals and oil from the desert to non-desert destinations. Firewood and charcoal as fuel for desert inhabitants is transported from non-desert areas through desert roads. Altogether, ground transport now provides for a significant transportation sector in deserts. Better road systems have also opened up for tourism, one of the fastest growing economic sectors. This modern transport goes through oases and sky-islands and facilitates the urbanization of major oases, thus contributing to the economy of desert regions but also presenting risks to traditional lifestyles. Desert roads, however, are vulnerable to flash-floods and drifting sand, and programmes directed toward the construction of physical wind barriers or establishing vegetation to reduce drifting sand exist (Figure 3.5).

The use of modern cross-desert roads is also often constrained by armed bandits and guerrilla warfare that may include road mining, making travel and trade dangerous and uncertain. Although no

Figure 3.5: Desert road protection

Desert roads are protected from moving sands with irrigated road-side plantations. Taklimakan Desert, China.
Source: Wang Tao

statistics are available, much of the cross-desert trade and road use is currently of an illegal nature — drugs, arms and slavery (mostly for prostitution); 80–90 per cent of the heroin consumed in Europe comes from the deserts of Afghanistan, and 60 percent of Afghan opium and heroin travels through Central Asian or West Asian deserts. Products from poaching of endangered species also travel through deserts (Nellemann 2005).

CROSS-DESERT ANIMAL MIGRATION
Trans-desert bird migration

Thousands of bird species are migratory, performing a north-south, often cross-equatorial, seasonal migration between northern and southern homes (Figure 3.6), of which the one north of the subtropical region is used for breeding. A migratory bird is an integral component of these two distinct ecosystems, linked only by migration routes, used for reaching its destination in the fastest and safest way. Since most deserts are wedged within the subtropical latitudes, since many regions within subtropical latitudes are deserts, and since competition for resources at either end favors early arrival, most migrants select the shortest route, which is very likely to include a desert-crossing section serving as a corridor for the migration of large numbers of birds.

Migrating is always costly, but more so on desert crossing, because migrants spend most of their life outside the desert and hence are not as well-

Figure 3.6: Major migratory bird routes of the world

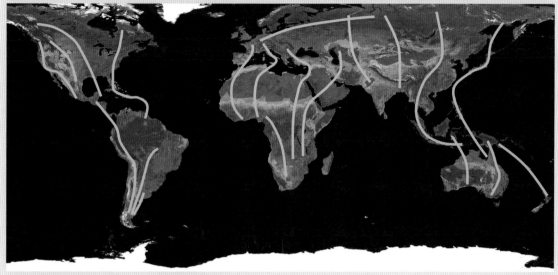

The red areas show the world's deserts.
Source: Perrins and Elphick 2003

adapted to deserts as resident birds. Yet, since crossing deserts rather than circumventing them evolved as an adaptation to reduce travel time, the added cost is reduced by specific adaptations. The fuel used for travel is fat—light in weight and rich in energy —stored under the skin. A 10g warbler needs fat stores approximately a *quarter* of its body mass to complete the flight across the Sahara Desert, which can take 40 hours (Carmi and others 1992). Burning this fat during the intensive and exhaustive flight generates much heat and the rapid breathing during flight removes body water. To reduce the risks of overheating and dehydration, most small cross-desert migrants fly at night. When the desert tract to be crossed is short, birds may make it in a single flight, but otherwise birds conduct intermittent flights between stopovers (Biebach and others 2000); they alight at dawn, seeking shaded and concealed refuges, even as small as a shading rock or a single bush if nothing better is in sight, where they minimize water loss and rest prior to taking off again when night falls.

Migrants can do even better when stopping-over in desert oases, which function as stepping stones for *en route* energy replenishment (Safriel and Lavee 1988). But because oases are rare (for example, the combined area of Saharan oases is only 2 per cent of this desert), scattered and hence difficult to spot, many migrants use routes along desert rivers such as the Nile (Box 3.1). Combined, all desert oases and riparian corridors comprise a critical habitat for a hugely disproportionate number of migratory species. Yet, the richer an oasis or a riparian tract is, the more it is likely to be impacted by people (two-thirds of the human population of the Sahara concentrate in oases) who can pose a threat to the migrants; the intensification of irrigated agriculture in natural and man-made oases, for example, may add open water and some locally abundant sources of food, which does attract migrants, but agrochemicals and other pollutants may harm the migrants, either directly, or through damage to the insects they might feed on (e.g., Evans and others 2005).

Cross-desert bird migration is also being impacted by global climate change. Cues such as day length, which are independent of climate, are involved in determining the onset of migration. These have evolved to synchronize with changes in seasonal abundance of plant and insect food resources in the two far-apart "homes" of the migratory birds; in contrast with the day-length cues, these changes in food availability are highly dependent on climate. Climate change will decouple the synchrony between non-climatic cues and climatic events. This synchrony is critical for cross-desert migrants, many of which time their migration such that they arrive at their feeding ground, just prior

The white stork (*Ciconia alba*) is a soaring, cross-desert daytime migrant, with a breeding home in Europe and a wintering one in southern Africa. The map describes the trajectory of "Princess", a female white stork fitted with a satellite-tracked transmitter. When at least nine years of age, Princess left her breeding grounds (red circle) in northern Germany (51.7°N) on 25 August 2002, arrived at the northern edge (the Negev in Israel, 31.5°N) after 15 days, then travelled south across the Sinai and along the Nile Valley. After exiting the Sahara Desert, thus ending a 9 days cross-desert journey, Princess stayed for 58 days in the southern part of the Nile basin, within the semi-arid area next to the desert edge (white rectangle) to replenish reserves spent on the first leg of the voyage and to store for the post-desert second leg towards her wintering home (white circle). She then travelled 20 days, more leisurely than the desert crossing, and arrived at her winter grounds, which included Zimbabwe, Botswana and eventually the Cape Province at the southernmost tip of the African continent.

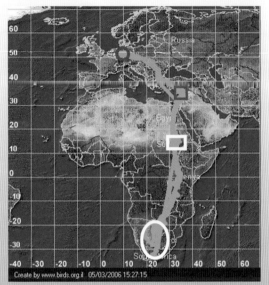

Princess wandered within her wintering grounds for 115 days of the southern summer. She started her journey back to Germany on 30 March 2003, travelling rather quickly and arriving at the desert edge five days later. Here, she again stayed for restocking, this time only for 11 days. The northbound desert crossing was longer than the southbound one, taking 15 days. Upon leaving the desert she stayed in the semi-arid area of Israel (red rectangle) for 15 days of replenishment following the desert crossing. It then took her 25 days to travel through Lebanon, Turkey and Europe until reaching her breeding home in northern Germany rather late, on 9 June, then spending only 65 days there, and again taking off on 14 August 2003.

Altogether, Princess covered 12 600 km in 102 days of her southbound migration, of which 2 440 km were through a desert, yet close to riparian habitats of the Nile. Her northbound voyage was shorter – 8 250 km in 71 days. Yet, her late arrival and short stay in the breeding grounds attest to an unsuccessful breeding. In the previous year she arrived on April 12, and stayed for 98 days, a period sufficient for completing successful reproduction.

Source: Migratory Birds Know No Boundaries, International Center for the Study of Bird Migration, Israel Ornithological Center, http://www.birds.org.il/show_item.asp?levelId=457

to entering, or just after exiting the desert when the food supplies there peak, which enables them to replenish their fat reserves for the remaining journey (Vickery and others, 1999). Compounding that problem is the projected "expansion" of deserts, which would increase the distance of that flight, perhaps beyond that which birds are already adapted to, or to which they might rapidly adapt. Furthermore, even if climate change does not jeopardize the voyage itself, it may reduce the benefits of migration. Since migrants have only a small margin of safety of energy reserves during migration, the condition in which they arrive at their destination, where they may encounter intense competition, substantially determines their survival.

To conclude, the cross-desert migration of birds, many of which are both familiar and important to people living far from deserts, is sensitive to human impact, climate-change included. Since this migratory network can only be as strong as its

weakest link, the conservation of desert sites used by alighting migrants, as well as off-desert ones on which the success of cross-desert migration depends, is urgently required (Hutto 2000).

Locusts moving through deserts

Though there are many more insect species than bird species, the number of migratory insects is smaller than that of migratory birds and only one small group of migratory insects is associated with deserts — the locust. Unlike migratory birds which cross deserts in a regular, seasonal two-way migration, locusts cross the desert in a unidirectional, irregular pattern. And, whereas the arrival of migratory birds is often welcome, the sighting of locust swarms is always ominous. Several locust species spend part of their lives in deserts and many locust swarms cross deserts. Most significant is the desert locust (*Schistocerca gregaria*) of the least dry parts of deserts — arid (but not hyperarid) regions in 25 countries of the Sahel (including Burkina Faso, Chad,

Mali, Mauritania and Niger), the Arabian Peninsula, along the coast of the Red Sea, and along the coast of the ROPME Sea Area (Regional Organization for the Protection of the Marine Environment — Kuwait Regional Convention, 1978) up to the India-Pakistan border (Simpson and others 1999). The desert locust in the solitary phase poses no threat to crops; its small populations are dispersed in patches of suitable habitat, with little movement between them (Ibrahim and others 2000).

When spells of good rainfall occur in several successive rainy seasons, the soil becomes moister and vegetation grows more quickly. In response the animals also grow more quickly and egg-laying in holes dug in the soft, moist soil intensifies. Each of the formerly isolated populations increases, and their movement downwind leads to a concentration of several crowded, fast-growing populations (Despland and others 2004). Once the aggregated population is large and crowded, the individuals change in colour, physiology and behaviour which, combined, helps them to aggregate and reproduce intensively (Pener 1991) especially when rains persist, thus increasing pressure on resources. When, eventually, vegetation is decimated at the source, a large-scale directional, downwind flight of whole swarms is initiated (Culmsee 2002).

This movement can bring the swarms to other desert arid or hyperarid areas, or to non-desert areas (Figure 3.7) — the "outbreak areas" (as distinct from the areas they inhabited prior to the massive movement — the "recession areas", Pedgley 1981). Carried by the wind which takes them to where it takes the rain too (Waloff 1960), swarms that may contain 50 000 million individuals, migrate from deserts into non-desert areas, where they can spread over more than 20 per cent of

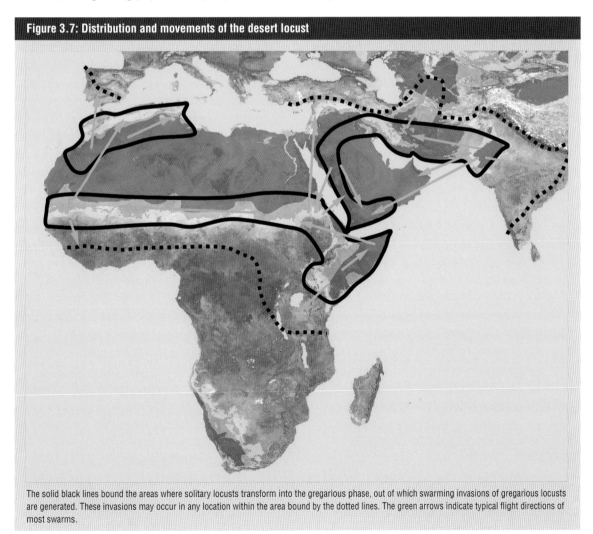

Figure 3.7: Distribution and movements of the desert locust

The solid black lines bound the areas where solitary locusts transform into the gregarious phase, out of which swarming invasions of gregarious locusts are generated. These invasions may occur in any location within the area bound by the dotted lines. The green arrows indicate typical flight directions of most swarms.

the land surface of up to 65 countries including in the Sahara and the Arabian Desert. They can reach southern Spain, Turkey, West Africa, India, Bangladesh, Tanzania and the Democratic Republic of Congo, consuming 100 000 tons of vegetation a day. Swarms originating in deserts can even cross oceans; during the 1986–89 plague, swarms escaped extermination in western Africa and crossed 5 000 km over the Atlantic Ocean, reaching the West Indies and the eastern coast of Venezuela. Swarms migrating to outbreak areas rarely return to the original recession areas (Ibrahim and others 2000): they die out, or are destroyed by cold weather or pest control measures. Since it is the desert edges that are often the source of locust plagues, it is there, in countries such as Algeria, Somalia, Sudan and Iran, where the battle against locusts can be most-effectively fought, in order to save crops in other areas, in countries such as Mali, Niger, Chad and Yemen (Showler 2002).

The Impact of Desert Research on Global Science

Though remote, isolated and of harsh climatic conditions, deserts attract scientists of every discipline. Most of them carry out scientific research on deserts, but some perform scientific research in deserts, capitalizing on some desert attributes that make the research there more productive than if carried out elsewhere. Both research on deserts, striving to generate knowledge of this specific environment, and research in deserts, which makes use of this environment for increasing general knowledge, have made an impact on global science. This is because both types of research carried out in deserts benefit from the desert's clean atmosphere, sparse human settlements, dry climate, sparse vegetation cover, and often thin soil cover — features that contribute to good preservation conditions, visibility, and high detectability of scientifically-relevant objects and phenomena.

RESEARCH IN DESERTS CONTRIBUTES TO SPACE EXPLORATION

Since the desert environment is most reminiscent of that of several barren, apparently lifeless planetary bodies, deserts have been used as simulators for testing planetary exploration equipment. For example, NASA conducted experiments in US southwestern deserts to test communication rovers and to improve human-robot interactions in conditions similar to those on the Moon and Mars (Volpe 1999). The Nomad robot, designed for long-distance planetary exploration, successfully traversed the Atacama Desert, and a remotely-guided rover explored the distribution and diversity of life in this desert as an analogue to Mars, not only because it is extremely dry, but also, because, like Mars, it experiences high levels of ultraviolet radiation, due to its altitude and atmospheric transparency (Wettergreen and Cabrol 2005).

Just as deserts are used for research that helps to send vehicles and sensors to explore distant planets, meteors arrive on earth from outer space; here deserts play a role too, serving as a repository of space debris and meteors reaching the planet's surface and remaining there as meteorites, well-preserved due to the slow rate of desert rock erosion. Indeed, most collected meteorites have come from deserts, though many are now also collected in Greenland and Antarctica.

Unlike the use of deserts for testing technologies for space exploration and for detecting meteorites, astronomical observations can theoretically be conducted in any environment, but conducting them in deserts is advantageous. This is because the background "noise" relative to the "signal" emitted by faint celestial sources is minimized in deserts, except where and when dust storms are often generated. This is because, beside the obvious low cloud cover, deserts minimize light pollution from human settlements, atmospheric water vapour, and air turbulence close to the telescope. It is easy to find desert sites where flat, treeless areas reduce the turbulence of airflow and in which urban encroachment is highly unlikely. Indeed, some of the largest and most expensive astronomical instruments of the international astronomical community are placed on desert mountain tops, where also the layers of atmosphere present between the telescope and any celestial object are minimized, such as, for example, the Very Large Telescope array (VLT) of the European Southern Observatory (ESO) on Cerro Paranal in the Atacama high desert of Chile, the Sutherland site in the Karoo of South Africa where the SALT telescope started operations in 2005, and other observatories currently being planned for other desert sites.

Deserts are treasures of paleontological findings

The world's deserts serve as a natural laboratory for investigating the history of life, be it that of plants and animals or of humankind. This is because the desert's sparse vegetative cover, lack of thick soils, and aridity combine to provide large areas of exposed rock. This, together with the scant precipitation that reduces chemical leaching by groundwater, promotes the preservation and the detection of fossils that allow deciphering the evolutionary history of animals and plants, and of early man (Figure 3.8). Although ultimately it is the distribution of appropriate aged sedimentary rocks that determines where fossils will (and will not) be discovered, deserts probably have produced a disproportionate number of major paleontological finds.

Figure 3.8: Desert fossils

A fossil ammonite shows-up in the sandy surface of the Chihuahuan Desert, at 1500 m altitude, as a silent record of past geologic ages and of the power of tectonic forces that brought the floor of the ancestral Atlantic Sea into the contemporary Chihuahuan Plateau.
Source: Patricio Robles-Gil

Particularly noteworthy examples include the badland terrains of the Gobi Desert where a great diversity of late Cretaceous (65 million year-old) dinosaurs and mammals have been unearthed; the Sahara's Ténéré Desert in Niger where excavations in lower Cretaceous — 10 Ma (million years ago) — rocks recently recovered over 25 tons of dinosaur fossils; the Karoo Desert of South Africa and its exposures of upper Permian (250 Ma) through lower Triassic (220 Ma) rocks containing abundant remains of mammal-like therapsid reptiles; Egypt's Fayum Desert, which preserves important Eocene-age (40 Ma) fossil treasure of early cetaceans and sirenians; the Pisco Basin, a coastal desert in Peru, where an exposed stratigraphic sequence of Miocene to Pliocene (20 to 2 Ma) rocks has produced spectacular assemblages of fossil marine mammals; and the Colorado desert in south-eastern California, that harbours one of the most complete records of late Cenozoic land mammal evolution in North America.

Findings in deserts shed light on the origins of mankind and its culture

Some fossils of hominids found in deserts in recent years were instrumental in reconstructing the evolution of humans. The early hominid *Sahelanthropus tchadensis*, a new genus and species of hominid that lived 6–7 million years ago (Vignaud and others 2002) was discovered in 2001 in the Saharan Djurab Desert of northern Chad. This finding suggests that the divergence between the human and chimpanzee lineages was earlier than indicated by most molecular studies. *Australopithecus garhi*, who lived 2.5 million years ago, was found in 1996–9 in the Afar Desert of Ethiopia. *Australopithecus garhi* might have been the world's earliest maker of stone tools, used to scrape bones of hunted mammals (Asfaw and others 1999). The oldest found fossilized remains of modern humans, who lived 160 000 years ago, was found in 1977 in the desert sands near the Ethiopian village of Herto (White and others 2003).

Most desert attributes that conserve fossils also apply to preservation of prehistoric and archaeological remains; and the low level of surface disturbance by humans has also preserved prehistoric sites. These attributes have resulted in a large number of archaeological discoveries in

Figure 3.9: Archeological remains in the Negev Desert

Aerial photo of an ancient threshing-floor 'Uvda Valley, southern Negev, Israel. The larger grey oval (18 × 36 m) was built on the rock surface in the 5th millennium BCE, while the smaller structure (14 × 17 m) was dug into the rock in the early 3rd millennium BCE. Silos and a flint workshop were built next to it.

Source: Uzi Avner

deserts, many of which are of global significance. For example, the earliest stone plough-tips, the largest and oldest cluster of threshing floors (Figure 3.9) and the earliest run-off irrigation system supporting large cultivated fields, belonging to early farming communities which evolved from hunter-gatherers since 6 000 BCE, were found in the Negev Desert (Avner 1998). Among hundreds of prehistoric cult sites in West Asian deserts, shrines of standing stones representing deities have been found (Figure 3.10), erected first in the 12th millennium BCE and becoming very common from 6 000 BCE. These suggest that while peoples

of the fertile lands of that region worshiped gods in figurative, human or animal forms, natural, unshaped stones represented gods to the desert people. Millennia later, this abstract, non-figurative theology was also adopted by the Jewish, Nabatean and Islamic religions (Avner 2000). Another rather famous desert finding with religious implications, that of the Dead Sea Scrolls, preserved for 2 000 years due to the desert's dry climate and discovered since 1947, shed new light and contributed insights into the history, philosophy and evolution of Judaism and early Christianity. Less widely known but of far-reaching impact is the desert origin of alphabetic writing, which first appeared in West Asian desert rock inscriptions around 2 000 BCE. While the ancient Egyptian and Mesopotamian scripts consisted of hundreds of complex signs (Figure 3.11), a group of people in the Sinai desert adopted only 28 (mostly Egyptian) symbols, each representing a single consonant. Later, these signs evolved into the Phoenician and the Hebrew script, then to the Greek and Latin and finally into the present-day Western European scripts.

Figure 3.10: Prehistoric desert shrines

A shrine with standing stones facing east, representing a group of seven deities, 5th–3rd millennia BCE, 'Uvda Valley', Negev Desert, Israel.

Source: Uzi Avner

Figure 3.11: Hieroglyphs and the development of alphabetic script

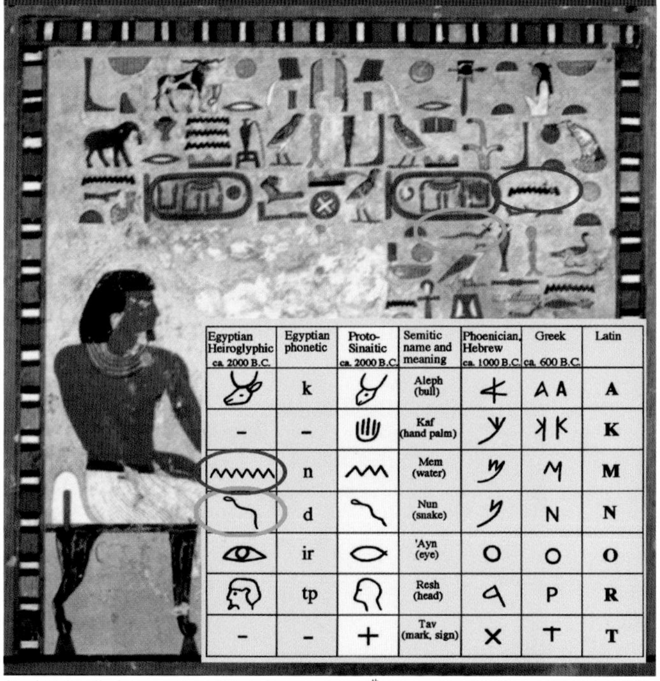

Egyptian Hieroglyphic ca. 2000 B.C.	Egyptian phonetic	Proto-Sinaitic ca. 2000 B.C.	Semitic name and meaning	Phoenician, Hebrew ca. 1000 B.C.	Greek ca. 600 B.C.	Latin
	k		Aleph (bull)		A A	A
–	–		Kaf (hand palm)		K	K
	n		Mem (water)		M	M
	d		Nun (snake)		N	N
	ir		'Ayn (eye)	O	O	O
	tp		Resh (head)		P	R
–	–	+	Tav (mark, sign)		T	T

A painted wall of the tomb of Prince Siremput II (12th Dynasty) in Aswan, Egypt. The encircled hieroglyphs are two of the five presented in the table inset. The table shows examples of proto-Sinaitic letters engraved by Semitic desert people around 2 000 BCE in rock inscriptions discovered in western Sinai.

Sources: Carpiceci 1997 (painted wall) and Yadin 1963 (table)

Desert research generated knowledge on the earth's geology

The low soil and vegetation cover of deserts has also attracted geologists and geomorphologists, whose research in deserts has contributed to our basic understanding of the processes that shape and mould the surface of the earth. The wealth of excellent rock exposures in the rocky deserts is outstanding, providing scientists and visitors with a glimpse into geological windows. Deep canyons, fault escarpments and rift valleys in deserts often exhibit sequences of rock strata that disclose chapters of hundreds-of-millions of years of earth history. Deserts have also been pivotal to the development of geomorphology, largely because their landforms are also easy to see. John Wesley Powell's report on the Grand Canyon (Powell 1875) opened geological science to the power of rivers, and Gilbert's (1877) work in North American deserts, are the foundational studies of modern geomorphology.

DESERT BIOLOGICAL RESEARCH CONTRIBUTES TO SEVERAL BIOLOGICAL SCIENCES

Desert research and the notion of convergent evolution

Being an environment as remote as imaginable from the aquatic origin of life, deserts attracted early life scientists eager to uncover the adaptations of desert organisms to this challenging setting. These endeavours had a profound impact on the disciplines of evolutionary biology, physiology and ecology. Regarding evolution, it was found that both desert annual plants and sedentary animals respond with extremely rapid growth to the short bouts of resource abundance, and by quiescence of life processes during the intervening long periods of shortage (Philippi 1993), whereas both perennial plants (Cabin and Marshal 2000) and mobile animals survive periods of low resource abundance by either moving to more favourable areas (animals), or by physiological means such as suppressing resource allocation to temporarily less important activities (plants). That groups with such different evolutionary ancestry as plants and animals display similar adaptations to the extreme and random fluctuations in resource availability convinced biologists that, in the living world, divergent genetic makeups have the potential to generate convergent solutions to a wide array of

environmental challenges and selection pressures (Smith and Wilson 2002) (see also "Biological Adaptations to Aridity" in Chapter 1).

Desert physiological adaptation as a model for life under stress

Desert research revealed that physiological adaptations enable mammals and birds to live in environments where water is limiting and temperatures are high, while human adaptations are exclusively behavioural and cultural. This prompted researchers to study the physiological responses of humans to desert conditions. These studies demonstrate the potential of people to acclimatize, rather than adapt, to stressful conditions (Shkolnik and others 1980). The fact that humans and their livestock do live in deserts has also compelled physiologists to examine the deleterious effects of high temperatures, chronic dehydration and food shortages on humans, and on the animals they domesticated due to their ability to live in deserts, such as camels (Schmidt-Nielsen and others 1967), goats, and donkeys (Izraely and others 1989).

What determines the number of links in a food chain? — A desert insight

Two features make deserts ideal for ecological research. The low plant cover enables one to easily explore animal activity either directly, or indirectly, by observing the tracks they leave on the bare soil surface. More importantly, since in deserts only one major factor, precipitation, governs ecological processes, and since the number of species in deserts is relatively low and the sizes of their populations are small, the desert ecosystem appears simpler, hence easier to understand than other ecosystems. These features encouraged the use of deserts as an outdoor laboratory, where hypotheses and theories developed in non-desert environments lend themselves to testing.

For example, a prevailing notion that evolving specializations for partitioning a limiting resource enables many species to avoid competition and coexist, thus leading to high diversity, is challenged by the finding that annual plants and darkling beetles (the blackish beetles of the family *Tenebrionidae*) exhibit high diversity in deserts, but subsist on resources not amenable for partitioning;

since soil moisture is restricted to its thin top layer, the coexistence of so many annual plant species cannot be attributed to each of them drawing water from a different depth. Similarly, the rather physically and chemically homogenous plant litter cannot be partitioned and is indiscriminately consumed by all darkling beetles. This desert observation supports the notion that species diversity can be maintained not only by competition that generates specialization, but also by predation (Ayal and others 2005).

This leads to challenging yet another central paradigm of current ecological theory, that food chain length is determined by the productivity of its first link, the primary productivity of plants, and that high primary productivity maintains long food chains. In deserts, however, long food chains with several predation links on top have been observed repeatedly, in spite of the desert's overall low primary productivity. Several related observations explain this finding. Most of the desert's primary productivity is not consumed by herbivores but becomes plant litter; plant litter in deserts is not readily decomposed by soil micro-organisms, which are constrained by the desert's protracted periods of low moisture. Hence, much litter accumulates on the surface and is consumed by a large number of arthropods. Being relatively small, these litter-consuming arthropods are preyed upon by only slightly larger small predators, such as arachnids and reptiles, which in turn are preyed upon by birds and mammals, which are larger still. Thus, desert food chains are long and size-structured, yet are supported by a base of low primary productivity (Ayal and others 2005). Desert research then implies that it is the body size of the primary consumers rather than the quantity of primary production that determines the length of food chains, a conclusion that undermines the high productivity–long food chain paradigm, and which may apply to other, non-desert ecosystems.

Why are linkages important?

Exploring how deserts are linked to the rest of the planet, this chapter highlights the importance of deserts. It also underscores why the more than 6 000 million people who live outside the desert biome need to take an interest in what happens in deserts, even though only 144 million people currently live there. This is not only because so

much of the oil and so many of the diamonds come from deserts; there are more subtle aspects of human culture that for inexplicable reasons have been nurtured in deserts, such as the advent of the alphabetic script or the emergence of monotheistic religions that, respectively, catalyzed human development and largely dominate human relations the world over. Yet, it is not the signature of a desert's remote past that matters most. Rather, much of human well-being in its broadest and most global sense depends in several ways on what happens in deserts today.

For example, the climate system of the areas beyond the desert affects that of the deserts themselves, but some of the desert climates' responses to these then affect the climate of the non-desert world. This in turn much depends on global climate change, mostly generated by non-desert people. Deserts may respond to these changes, among other things, by increased emissions of cross-boundary dust storms with far-reaching negative (as well as positive) implications. Another example is derived from the dependence of non-desert birds on cross-desert migration. Birds are directly, but people are indirectly, affected, because when not on the move these birds are intimately involved in the provision of services in the non-desert ecosystems in which they live, services that support life in general and human well-being in particular, at local and global scales, and which will not be provided if the cross-desert migration is impaired. Hence, this migration depends on what people both in deserts and outside of deserts do, either to protect or to disrupt these trans-desert migrations. Furthermore, these two groups of people, the desert and the non-desert, are inter-connected too. The livelihoods of many desert people, upon which the flow of benefits from deserts to the rest of the world depends, is often linked to the ways non-desert people manage the non-desert headwaters of major rivers that cross into deserts, nourish their life and nurture their societies.

To conclude, our understanding of global processes, the development of much of our modern research, our ability to cope with global environmental change, and the preservation of much of our global heritage depends to a large extent on the way we manage and preserve the world's deserts. What happens in deserts affects every one of us, wherever we are. What happens outside deserts impacts deserts, changes the way they function, and what they can contribute to the rest of the planet.

REFERENCES

Ahmad, H., Bhatti, G.R., and Latif, A. (2004). Medicinal flora of the Thar Desert: An overview of problems and their feasible solutions. *Zonas Áridas* 8: 74–84

Asfaw, B., White, T., Lovejoy, O., Latimer, B., Simpson, S., and Suwa, G. (1999). *Australopithecus garhi*: A New Species of Early Hominid from Ethiopia. *Science* 23: 629–635

Avner, U. (1998). Settlement, Agriculture and Paleoclimate in Uvda Valley, Southern Negev Desert, 6th–3rd Millennia B.C. In *Water, Environment and Society in Times of Climate Change* (eds. A. Issar and N. Brown) pp. 147–202. Dordrecht, Netherlands

Avner, U. (2000). Nabatean Standing Stones and Their Interpretation. *ARAM* 11–12, 97–122

Ayal, Y., Polis, G.A., Lubin, Y., and Goldberg, D.E. (2005). How can high animal biodiversity be supported in low productivity deserts? The role of macrodetrivory and physiognomy. In *Biodiversity in Drylands* (eds. M.Shachak and R.Wade). Cambridge University Press, Cambridge

Batjes, N.H. (1996). Total carbon and nitrogen in the soils of the world. *European Journal of Soil Science* 47: 151-163

Batjes, N.H., and Sombroek, W.G. (1997). Possibilities for carbon sequestration in tropical and subtropical soils. *Global Change Biology* 3: 161–173

BGS (2006). *World Mineral Production 2000–2004*. British Geological Survey, Keyworth, Nottingham

Biebach, H, Biebach, I., Friedrich, W., Heine, G., Partecke, J., and Schmidl, D. (2000). Strategies of passerine migration across the Mediterranean Sea and the Sahara Desert: a radar study. *Ibis* 142: 623–634

Cabin, R.J., and Marshall, D.L. (2000). The demographic role of soil seed banks. I. Spatial and temporal comparisons of below- and above-ground populations of the desert mustard *Lesquerella fendleri*. *Journal of Ecology* 88: 283–292

Carmi, N., Pinshow, B., Porter, W.P., and Jaeger, J. (1992). Water and energy limitations on flight duration in small migrating birds. *Auk* 109: 268–276

Carpiceci, A.C. (1997). *Art and History of Egypt: 5000 Years of Civilization*, Bonechi Books, Florence, Italy

Charney, J.G. (1975). Dynamics of deserts and drought in the Sahel. *Quarterly Journal of the Royal Meteorological Society* 101: 193–202

Conway, D. (2005). From headwater tributaries to international river: observing and adapting to climate variability and change in the Nile basin. *Global Environmental Change* 15: 99-114

Crawford, L. (1990). Amax Pulling Out of Lithium Project. *Financial Times*, pp. 36, 8 November 1990

Culmsee, H. (2002). The habitat functions of vegetation in relation to the behaviour of the desert locust *Schistocerca gregaria* (Forskål) (Acrididae: Orthoptera) — a study in Mauritania (West Africa). *Phytocoenologia* 32: 645–664

Degens, E. T., Kempe, S., and Richey, J. (eds.) (1990). *Summary: Biogeochemistry of Major World Rivers*. In *Biogeochemistry of Major World Rivers*. SCOPE Report 42, John Wiley & Sons, Chichester (available online at www.icsu-scope.org)

Despland, E., Rosenberg, J., and Simpson, S.J. (2004). Landscape structure and locust swarming: a satellite's eye view. *Ecography* 27: 381–391

Divon H., and Abou-Hadab, F. (1996). *Collaborative Agricultural and Rural Development of Settlement in the Nubariya Region, Egypt*. Jewish Virtual Library. http://www.jewishvirtuallibrary.org/jsource/Peace/egmashav.html [Accessed 19 April 2006]

Donaldson, J.R., and Cates, R.G. (2004). Screening for anticancer agents from Sonoran Desert plants: A Chemical Ecology Approach. *Pharmaceutical Biology* 42: 478–487

Donaldson, J.R., Warner, S.L., Cates, R.G. and Young, D.G. (2005). Assessment of antimicrobial activity of fourteen essential oils when using dilution and diffusion methods. *Pharmaceutical Biology* 43: 1–9

Evans, M., Amr, Z., and Al-Oran, R.M. (2005). The status of birds in the proposed Rum Wildlife Reserve, Southern Jordan. *Turkish Journal of Zoology* 29: 17–26

Gilbert, G.K. (1877). *Report on the geology of the Henry Mountains*. United States Department of the Interior, United States Government Printing Office, Washington, D.C

Golan-Goldhirsh, A., Sathiyamoorthy, P., Lugasi-Evgi, H., Pollack, Y., and Gopas, J. (2000). Biotechnological potential of Israeli desert plants of the Negev. *Acta Hort.* 523: 29–37

Golany, G. (1978). Planning Urban Sites in Arid Zones: The Basic Considerations. In *Urban Planning for Arid Zones* (ed. G. Golany) pp. 3–21. John Wiley & Sons, New York.

Gyan, K., Henry, W., Lacaille, S., Laloo, A., Lamsee-Ebanks, C., McKay, S., Antoine, R.M., and Monteil, M.A., (2005). African dust clouds are associated with increased paediatric asthma accident and emergency admissions on the Caribbean island of Trinidad. *International Journal of Biometeorology* 49: 371–376

Hanspeter, L., Weingartner, R., and Grosjean, M. (1998). *Mountains of the World: Water Towers for the 21st Century*. Institute of Geography, University of Berne and Swiss Agency for Development and Cooperation (available online: Mountains of the World website www.mtnforum.org/resources/library/magen98a2.htm)

Hutto, R. L. (2000). On the importance of *en route* periods to the conservation of migratory land birds. *Studies in Avian Biology* 20: 109–114

Ibrahim, K.M., Sourrouille, P., and Hewitt, G.M. (2000). Are recession populations of the desert locust (*Schistocerca gregaria*) remnants of past swarms? *Molecular Ecology* 9: 783–791

IEA. (2005). *World Energy Outlook 2005 — Middle East and North Africa Insights*. International Energy Agency, Paris

IPCC (2001). "Climate Change 2001. Working Group I, Third Assessment Report". Cambridge University Press, Cambridge

Izraely, H., Choshniak, I., Stevens, C.E., and Shkolnik, A. (1989). Energy digestion and nitrogen economy of the domesticated donkey (*Equus asinus*) in relation to food quality. *Journal of Arid Environments* 17: 97–101

Jianlin, H., Ochieng, J.W., Lkhagva, B., and Hanotte, O. (2004). Genetic diversity and relationship of domestic Bactrian camels (*Camelus bactrianus*) in China and Mongolia. *Journal of Camel practice and Research* 11: 97–99

Jickells, T. D, An, Z.S., Andersen, K.K., Baker, A.R., Bergametti, G., Brooks, N., Cao, J.J., Boyd, P.W., Duce, R.A., Hunter, K.A., Kawahata, H., Kubilay, N., LaRoche, J., Liss, P.S., Mahowald, N.M., Prospero, J.M., Ridgwell, A., Tegen, I., and Torres, R. (2005). Global iron connections between desert dust, ocean biogeochemistry, and climate. *Science* 308: 67–71

Kishcha, P., Alpert, P., Barkan, J., Kirchner, I., and Machenhauer, B. (2003). Atmospheric response to Saharan dust deduced from ECMWF reanalysis (ERA) temperature increments. *Tellus* B, 55: 901–913

Koocheki, A., and Nadjafi, F. (2003). *The status of medicinal and aromatic plants in Iran and strategies for sustainable utilization* Medicinal Plants Network. http://www.medplant.net/modules/DownloadsPlus/uploads/Publications/Nadjafi_Paper_on_Medicinal_Plants.doc [Accessed 19 April 2006]

Lal, R. (2002). Carbon sequestration in dryland ecosystems of West Asia and North Africa. *Land Degradation and Development* 13: 45–49

Lioubimtseva, E., and Adams, J.M. (2004). Possible implications of increased carbon dioxide levels and climate change for desert ecosystems. *Environmental Management* 33 (Supplement 1), S388–S404

Mahowald, N., and Luo, C. (2003). A less dusty future? *Geophysical Research Letters* 30(17): 1903

Masonen, P. (1997). Trans-Saharan Trade and the West African Discovery of the Mediterranean World. From: Sabour and Vikør, *Ethnic encounter and culture change*, pp.116–42. Bergen/London 1997

Monger, H.C., and Martínez-Ríos, J.J. (2000). Inorganic carbon sequestration in grazing lands. In *The Potential of U.S. Grazing Lands to Sequester Carbon and Mitigate the Greenhouse Effect* (eds. R.F. Follett, J. Kimble and R. Lal) pp. 87–118. Lewis Publishers, Boca Raton, Florida

Morgan, J.A., Lecain, D.R., Mosier, A.R. and Milchunas, D.G. (2001). Elevated CO_2 enhances water relations and productivity and affects gas exchange in C3 and C4 grasses of the Colorado shortgrass steppe. *Global Change Biology* 7: 451

Naumberg, E., Housman, D.C., Huxman, T.E., Charlet, T., Loik, M.E., and Smith, S.D. (2003). Photosynthetic responses of Mojave Desert shrubs to free air CO_2 enrichment are greatest during wet years. *Global Change Biology* 9: 276–285

Nellemann, C. (ed.) (2005). *The fall of the water: Emerging threats to the water resources and biodiversity at the roof of the world to Asia's lowland from land-use changes associated with large-scale settlement and piecemeal development.* UNEP/GRID-Arendal, Arendal, Norway (available online at www.globio.info)

Nianthi, K.W.G.R., and Husain, Z. (2004). Impact of climate change on rivers with special reference to river-linking project. In *Regional Cooperation on Transboundary Rivers: Impact of the Indian River-linking Project,* (eds. M.F. Ahman, Q.K. Ahmad, and M. Khalequzzaman) pp. 263–278. Bangladesh Environment Network. http://www.ben-center.org/ConfPapers-2005/Rekha,doc [Accessed 19 April 2006]

Okin, G.S., Mahowald, N., Chadwick, O.A., and Artaxo, P. (2004). Impact of desert dust on the biogeochemistry of phosphorus in terrestrial ecosystems. *Global Biogeochemical Cycles*, 18, Art. No. GB2005

Pedersen, J. (1995). Drought, migration and population growth in the Sahel: the case of the Malian Gourma: 1900–1991. *Population Studies* 49: 111–126.

Pedgley, D. (1981). *Desert Locust Forecasting Manual.* Centre for Overseas Pest Research, London

Pener, M.P. (1991). Locust phase polymorphism and its endocrine relations. *Advances in Insect Physiology* 23: 1–79

Perrins, C.M., and Elphick, J. (2003). *The complete encyclopedia of birds and bird migration.* Chartwell Books, Inc., Edison, N.J

Peterson, A.T., Ortega-Huerta, M.A., Bartley, J., Sánchez-Cordero, V., Soberón, J., Buddemeier, R.H., and Stockwell, D.R.B. (2002). Future projections for Mexican faunas under global climate change scenarios. *Nature* 416: 626–629

Philippi, T. (1993). Bet-hedging germination of desert annuals – beyond the 1st year. *American Naturalist* 142(3): 474–487

Pinty, B., Roveda, F., Verstraete, M.M., Gobron, N., Govaerts, Y., Martonchik, J.V., Diner, D., and R. Kahn (2000). Surface Albedo Retrieval from METEOSAT. *Journal of Geophysical Research* 105: 99–134

Plant Talk On-Line (1997). *Plants in traditional and herbal medicine.* http://www.plant-talk.org/Pages/PFacts10.html [Accessed 19 April 2006]

Portnov, B.A., and Erell, E. (1998). Long-term Development Peculiarities of Peripheral Desert Settlements: The Case of Israel, *International Journal of Urban and Regional Research* 22(2): 216–32

Powell, J.W. (1875). *Exploration of the Colorado River of the West (1869-72).* Reprinted by University of Chicago Press, Washington, D.C

Prospero, J. M., Ginoux, P., Torres, O., Nicholson, S.E., and Gill, T.E. (2002). Environmental characterization of global sources of atmospheric soil dust identified with the NIMBUS 7 TOMS absorbing aerosol product. *Reviews of Geophysics* 40, Art. No. 1002

Richardson, C.J., Reiss, P., Hussain, N.A., Alwash, A.J., and Pool, D.J. (2005). The Restoration Potential of the Mesopotamian Marshes of Iraq. *Science* 307: 1307–1311

Rosenfeld D., Rudich, Y., and Lahav, R. (2001). Desert dust suppressing precipitation — a possible desertification feedback loop. *Proceedings of the National Academy of Sciences* 98: 5975–5980

Safriel, U., and Lavee, D. (1988). Weight changes of cross-desert migrants at an oasis — Do energetic considerations alone determine the length of a stopover? *Oecologia* 76: 611–619

Schlesinger, W.H., (1985). The formation of caliche in soils of the Mojave Desert, California. *Geochimica et Cosmochimica Acta* 49: 57-66

Schlesinger, W.H., Reynolds, J.F., Cunningham, G.L., Huenneke, L.F., Jarrell, W.M., Virginia, R.A., and Whitford, W.G. (1990). Biological feedbacks in global desertification. *Science* 247: 1043-1048

Schmidt-Nielsen, K., Crawford, E.C., Jr, Newsome, A.E., Rawson, K.S., and Hammel, H.T. (1967). Metabolic rate of camels: effect of body temperature and dehydration. *American Journal of Physiology* 212: 341–346

Shinn, E.A., Smith, G.W., Prospero, J.M., Betzer, P., Hayes, M.L., Garrison, V., and Barber, R.T. (2000). African dust and the demise of Caribbean coral reefs. *Geophysical Research Letters* 27: 3029–3032

Shkolnik, A., Taylor, C. R., Finch, V., and Borut, A. (1980). Why do Bedouins wear black robes in hot deserts? *Nature* 283: 373–375

Showler, A.T. (2002). A summary of control strategies for the desert locust, *Schistocerca gregaria* (Forskål). *Agriculture, Ecosystems and Environment* 90(1): 97–103

Simpson, S.J., McCaffery, A.R., and Hägele, B.F. (1999). A behavioural analysis of phase change in the desert locust. *Biology Reviews* 74: 461–480

Smith, B., and Wilson, J.B. (2002). Community convergence: Ecological and evolutionary. *Folia Geobotanica* 37, 171–183

Smith, S.D., Huxman, T.E., Zitzer, S.F., Charlet, T.M., Housman, D.C., Coleman, J.S., Fenstermaker, L.K., Seemann, J.R., and Nowak, R.S. (2000). Elevated CO_2 increases productivity and invasive species success in an arid ecosystem. *Nature* 408: 79–82

Sokolik, I.N., Winker, D., Bergametti, G., Gillette, D.A., Carmichael, G., Kaufman, Y., Gomes, L., Schuetz, L., and Penner, J.E. (2001). Introduction to special section: Outstanding problems in quantifying the radiative impacts of mineral dust. *Journal of Geophysical Research* 106: 18015–18027

Tegen, I., Werner, M., Harrison, S. P., and Kohfeld, K. E. (2004). Relative importance of climate and land use in determining present and future global soil dust emission. *Geophysical Research Letters* 31

Tsoñis, A.A., Elsner, J.B., Hunt, A.G., and Jagger, T.H. (2005). Unfolding the relation between global temperature and ENSO. *Geophysical Research Letters* 32

Venkatesh, G. (2003). The Aussie mining sector. *Minerals and Metals Review* 29: 121–134

Vignaud, P., Duringer, P., Mackaye, H.T., Likius, A., Blondel, C., Jean-Boisserie, R., De Bonis, L., Eisenmann, V., Etienne, M-E., Geraads, D., Guy F., Lehmann, T., Lihoreau, F., Lopez-Martinez, N., Mourer-Chauviré, C., Otero, O., Rage, J-C., Schuster, S., Viriot, L., Zazzo, A., and Brunet, M. (2002). Geology and palaeontology of the Upper Miocene Toros-Menalla hominid locality, Chad. *Nature* 418: 152–155

Volpe, R. (1999). Navigation Results from Desert Field Tests of the Rocky 7 Mars Rover Prototype. *International Journal of Robotics Research, Special Issue on Field and Service Robots* 18(7), July 1999

Walker, A.S. (1997). *Deserts: Geology and Resources.* United States Geological Service. http://pubs.usgs.gov/gip/deserts/ [Accessed 19 April 2006]

Waloff, Z.V. (1960). The fluctuating distributions of the desert locust in relation to the strategy of control. *Report of the 7th Commonwealth Entomological Conference*, London 6–15 July 1960, pp. 132–139 (available online at ispi-lit.cirad.fr/text/Waloff60a.htm)

Wettergreen, D., and Cabrol, N. (2005). Investigation of Life in the Atacama Desert by Astrobiology Rover. *EOS Transactions,* American Geophysical Union, 86(52), Fall Meeting. Supplement, Abstract P41D-04

White, T. D., Asfaw, B., DeGusta, D., Gilbert, H., Richard, G.D., Suwa, G., and Howell, C.H. (2003). Pleistocene *Homo sapiens* from Middle Awash, Ethiopia. *Nature* 423: 742–746

Yadin, Y. (1963). *The Art of Warfare in Biblical Lands*, McGraw-Hill, Toronto

WRI (2003). *EarthTrends.* World Resources Institute, Washington D.C

Yao, Y.G., Kong, Q.P., Wang, C.Y., Zhu C.L., and Zhang Y.P. (2004). Different matrilineal contributions to genetic structure of ethnic groups in the Silk Road region in China. *Molecular Biology and Evolution* 21(12): 2265–2280

Chapter 4: State and Trends of the World's Deserts

Lead authors: Stella Navone and Elena Abraham
Contributing authors: Marta Bargiela, David Dent, Cecilia Espoz-Alsina,
Alejandro Maggi, Elma Montana, Scott Morrison, Gabriela Pastor,
Hector Rosatto, Mario Salomón, Dario Soria, Laura Torres, Fidel Antonio Roig
Sergio Roig-Juñent, Clara Movia, Walter Massad

Oasis in Djado, Niger.
Source: Silvio Fiore/Topham Picturepoint

Since the Ordovician Period, 500 million years ago, when photosynthetic organisms first left the water to start colonizing the land, the history of plants has been that of a long fight against drought and the search for better strategies for the use of water. The final conquest of the land came some 300 million years ago, with the reduction of the male gametophyte to the pollen grain and the female gametophyte to the seed (during the Carboniferous Period, male gametes depended on water to reach the female gametes). This was enhanced by the development of angiosperms during the Cretaceous Period, 100 million years ago, when flowers evolved and insects and birds started acting as pollinators, and seeds appeared with clear adaptations for animal

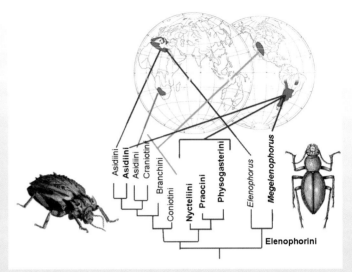

The widespread desert distribution and the geographically-entangled phylogenetic relationships of the *Pimelinae* (an ancient tribe of desert-dwelling *Tenebrionid* beetles) suggests that some desert-adapted organisms are evolutionarily very old, pre-existing the current distribution of deserts in the world.

Source: Flores & Roig-Juñent 2002

dispersal. This was the starting point of a long process of adaptation of plants and animals to symbiosis, which allowed plants to specialize, colonizing hostile environments. As a result, modern deserts are presently inhabited mostly by angiosperms.

In evolutionary terms, part of the desert biota is indeed very old. Some 200 million years ago, before the raising of the angiosperms, a group of gymnosperm plants known as Gnetales had already colonized many desert environments. One of their current descendants, the *Ephedra* species, is found in almost all deserts of the world (Archangelsky 1970). Other widespread desert taxa, like the creosote-bush family (Zygophyllaceae) or some tribes of Tenebrionid beetles, are also widespread in the deserts of the world, suggesting again the existence of desert-adapted groups before the breaking of Gondwana and Laurasia into the current continents, some 100 million years ago (Shmida 1985).

Apart from these disjunct desert taxa, the age of some deserts is also noticeable in their richness of endemic species. The older deserts, such as the Namib, are stable environments with xeric conditions that enabled the evolution of unique sets of highly endemic biodiversity, and, not surprisingly, its organisms show advanced adaptation to aridity.

The significant evolution that took place in the Cenozoic Period, after the continental break-up, is reflected in the high diversity of plants and animals that are now found restricted to specific desert areas. For example, the families Agavaceae (13 genera and 210 recent species) and Cactaceae (87 genera and 2 000 recent species) in the American deserts, and their complete absence in the Old World deserts, bears witness to rapid evolution in desert conditions in the Americas, possibly some 50 million years ago, well after the separation of the ancient continents (Pascual and Ortiz-Jaureguizar 1990). Similarly, the high richness of endemic species in the Namib Desert (including another relict species of the Gnetales, *Welwitschia mirabilis*, a remarkable paleo-endemism that bears witness to the antiquity of this desert), coupled with our knowledge of continental drift, suggests that the Namib has been arid for at least 55 million years and that the region, isolated between the ocean and the West African escarpment, has been a constant island of desert-like conditions in spite of worldwide climatic changes (Armstrong 1990).

In present-day deserts, local species richness — the number of species coexisting within a small local area — is very low, especially when compared to the dense aggregation of species observed in moist tropical areas. However, the mosaic heterogeneity of deserts — the way habitats and species replace each other in space — and endemism — the uniqueness of species living exclusively in an area and nowhere else — is extremely high in deserts. Thus, although deserts may be locally species-poor, their species are frequently of high conservation value, as they exhibit very restricted distributions. Additionally, the regional biological diversity of deserts may be unexpectedly high, because deserts are patchy and heterogeneous environments, with species closely adapted to dunes, slopes, rocky outcrops, ephemeral rivers, or salty flats. This results in a complex mosaic of narrowly-distributed species replacing each other along short-distance gradients and jointly presenting a rich array of locally-specialized plants and animals.

Additionally, many biological groups have traits that predispose their adaptation to arid environments, and as a result they have evolved high biological diversity in deserts. Arthropods in general, and insects in particular, have diversified extraordinarily in deserts, where they play important roles as major food sources in other animals' food cycles, as pollinators, as seed dispersers, and as decomposers of organic matter (Costello 1972). Within the vertebrates, reptiles are probably the most characteristic and conspicuous group in deserts (Jaeger 1957). Their scaly skin and their drought-avoiding behaviours make them resistant to dehydration. Rodents also show great diversity in deserts, with an amazing variety of sizes and forms, behaviours, and feeding habits. Within the plants, grasses, and composites, some succulent families such as cacti, aloes, and euphorbia are especially abundant in deserts.

Finally, relative to those in non-desert environments, species in deserts tend to be less related to the other species with which they occur. In other words, a given number of species in deserts will likely represent a greater diversity of higher-order taxa (such as genera or families) than a similar number in non-deserts. This implies that the relatively low species richness of deserts is partially compensated by the occurrence of many taxonomically unrelated groups, with widely different life-forms, coexisting side by side. This factor adds value to the uniqueness of the species guilds and communities of deserts. In the North American deserts, for example, the number of species per genus is 4–5, while in non-desert habitats the value is 10 (Stebbins 1974). In the Atacama desert of Chile, the mean number of species per genus is even lower (1–2), a fact that highlights the astounding number of contrasting life-forms of different evolutionary origin coexisting in the field. Apart from its influence in higher-order diversity, this phenomenon also has an impact on the wide spectrum of life-forms that is found in deserts: cactoid succulents, long-lived woody trees, small-leaved shrubs, dune creepers, bulb perennials, and seed ephemerals — to name but a few plant forms — forming heterogeneous communities.

In short, in spite of their low local diversity, deserts harbour a surprisingly high biological richness generated by their high biological heterogeneity in space (species turnover), high levels of endemism, high biological specialization in some taxa, and large diversity of higher-order taxa such as genera or families.

Box authors: Fidel Antonio Roig and Sergio Roig Juñent

Bound by the vegetation of the Indus Valley to the northwest, the Himalayas to the northeast, and the summer-rain tropics of India to the southeast, the Thar Desert covers 238 000 km² along the India–Pakistan border.
Source: NASA's "Blue Marble" website, http://earthobservatory.nasa.gov

density of 151 persons per square kilometre, and a human footprint of 33. These deserts receive 200 and 100 mm of annual rainfall respectively, enough to support trees, shrubs and grasses, in particular *Halostachys caspica* on salty soils and *Aristida pennata* on sandy soils. The rural population depends mostly on raising sheep and goats. About 63 per cent of the Indus Valley and 17 per cent of the Thar Desert are protected (the latter in 11 different locations).

NEARCTIC DESERTS
The Nearctic deserts are formed by five lowland deserts (Chihuahuan Desert, Sonoran Desert, Mojave Desert, Great Basin shrub steppe, and Meseta Central matorral); two coastal deserts (Baja California Desert and Gulf of California xeric scrub), and four montane relict sky-islands (Western Madrean Archipelago, Eastern Madrean Archipelago, Great Basin montane forests, and Sierra de Juárez and San Pedro Mártir pine-oak forests). They cover in total 1.7 million square kilometres, of which 19 per cent is under some level of legal protection. Because of the growth of large urban conglomerates such as Phoenix

in the US, their mean population density is high (44 persons per square kilometre), and their mean human footprint (21) is the second highest of the world's deserts, with footprint especially high in the Sonoran and Chihuahuan Deserts.

These deserts present a combination of sage brush (*Artemisia tridentata*) and creosote bush (*Larrea tridentata*) shrublands, halophytes (such as *Atriplex confertifolia*), and open grasslands in the moister parts. There are numerous cactus species except in the Great Basin shrub steppe, where freezing temperatures limit cactus growth. The giant saguaro cactus is emblematic of the Sonoran Desert; it grows very slowly but may reach a height of 15 m over its long life span of more than 200 years. The Baja California desert is home to over 3 500 species of endemic plants, almost a quarter of which are endemic, ranging from a diversity of shapes and sizes of cactus, to thick-stemmed trees and shrubs in the rocky mountain soils. Likewise, among the 3 200 plant species in the Chihuahuan Desert, around 1 000 are endemic. Creosote bush, tarbush, viscid acacia, yucca, and cactus are characteristic. The Joshua tree (*Yucca brevifolia*, Figure 4.4) is probably the most recognizable plant of the Mojave; growing with a wide variety of cactus, creosote bush, white bursage (*Ambrosia dumosa*), jojoba (*Simmondsia chinensis*), and small trees such as paloverde (*Parkinsonia microphylla*) and ironwood (*Olneya tesota*). Grazing, hunting and salt extraction (in the

Joshua tree (*Yucca brevifolia*) in the Mojave Desert.
Source: Patricio Robles-Gil

Baja California desert) are significant activities in these ecoregions, which also contain several large and fast-growing cities: Las Vegas, Reno, and Salt Lake City in the Great Basin, Los Angeles sprawling into the Mojave, Phoenix and Tucson in the Sonoran Desert.

NEOTROPIC DESERTS

The Neotropic deserts comprise three continental deserts (Low Monte, High Monte, and Central Andean Dry Puna), and two coastal deserts (Atacama and Sechura deserts, Figure 4.5). They cover 1.1 million square kilometres, of which only 6 per cent receives legal protection. Their mean population density is 18 persons per square kilometre, and their mean human footprint (16) is lower than in their North American counterparts,

Figure 4.5: Neotropical deserts

The Atacama desert is clearly visible in this satellite image of South America running along the Pacific coast of Chile and Peru. Cloud condensation above the cold waters of the Humboldt Current is seen entering the coast of southern Peru in the form of the *camanchaca* or morning fog. Great saltflats (*salares*) produced by intense evaporation are also discernible on the high altitude deserts of the Dry Central Puna.
Source: NASA's "Blue Marble" website, http://earthobservatory.nasa.gov

with most pressure concentrating in the Sechura Desert in the coasts of Peru. These deserts form a long arid continuum that cover South America's "arid diagonal", starting from the Pacific Ocean just south of the equator in Peru, and running in a southeast direction to the Atlantic coast of northern Patagonia, at latitude 43°S.

The barren landscape of the Atacama Desert features one of the driest deserts on earth. The almost complete absence of vegetation in its

interior is due to the lack of precipitation and the high mineral content of the soils. Rare rainfall events cause ephemeral plants to germinate and burst with flowers for a short period of time. The Monte, east of the Andes, is a fold desert with sandy plains, plateaus, and rocky foothills with an open, low thorn-scrub harbouring a characteristic endemic flora of zygophylls (family Zygophyllaceae). There are also edaphic communities of many species such as *Prosopis* thickets in ravines, shrub lands of broom rape (*Baccharis*) and saltbush (*Atriplex*) on clay soils, and *Allenrolfea vaginata* and *Suaeda divaricata* in salty soils. The dry Central Andean Puna carries tall tussocks of bunchgrass and other high-altitude grasses, shrubs like *Parasthrephia lepydophilla* and *Baccharis*, and a unique flora of high-altitude cushion plants. The main land use in the Puna is grazing with llamas, alpacas, goats and sheep. In ancient, pre-Hispanic times, these deserts were an important part of the Inca Empire; a network of roads stretched through them forming the *Qhapaq ñan*, or *Camino del Inca* (the Inca Road).

PALEARCTIC DESERTS

By far the largest set of deserts in the world is found in the Palearctic realm. The region includes 21 lowland desert ecoregions that cover an immense corridor of deserts stretching from the Atlantic coasts of Morocco, to the Mediterranean coasts of the Sahara, to the Gobi in Central Asia, including the deserts of northern Africa, Arabia, Azerbaijan, Taklimakan, Central Persia and the Caspian lowlands (the full list of ecoregions includes Alashan Plateau semi-desert, Arabian desert and East Sahero-Arabian xeric shrublands, Azerbaijan shrub desert and steppe, Badghyz and Karabil semi-desert, Caspian lowland desert, Central Asian northern desert, Central Asian southern desert, Central Persian desert basins, Eastern Gobi desert steppe, Gobi Lakes Valley desert steppe, Great Lakes Basin desert steppe, Junggar Basin semi-desert, Mesopotamian shrub desert, North Saharan steppe and woodlands, Qaidam Basin semi-desert, Red Sea Nubo-Sindian tropical desert and semi-desert, Registan-North Pakistan sandy desert, Sahara desert, South Iran Nubo-Sindian desert and semi-desert, South Saharan steppe and woodlands, and Taklimakan desert). The Palearctic realm also harbours three coastal deserts (Atlantic coastal desert, Gulf desert, and Red Sea coastal desert), as well as five montane sky-island relict ecosystems (Tarim Basin deciduous forests and steppe, Kuh Rud and Eastern Iran montane woodlands, Afghan Mountains semi desert, Tibesti-Jebel Uweinat montane xeric woodlands, and West Saharan montane xeric woodlands). The Palearctic deserts cover a remarkable 16 million square kilometres, totalling 63 per cent of all the deserts on the planet. Of this area, 9 per cent is legally protected. Their population density is 16 persons per square kilometre, and their mean human footprint (15) is the second lowest on the planet, possibly because of the sheer inaccessibility and the extreme aridity of many of its large ecoregions.

Figure 4.7: Palearctic deserts — Central Asia

Desert children of Kyrgyzstan, Central Asia.
Source: Patricio Robles-Gil

Sahara Desert is not well protected legally, but its inaccessibility plays a major role in conserving its ecosystems. The nomadic population depends on pastoral activities, hunting, and trade. Most settled people in the desert fringes do not venture into the interior. The high mountains shelter wild ancestors of many trees that have been domesticated for their fruits and nuts, such as pistachio and wild olive. Considering the hyper-arid conditions, the fauna of the Sahara is relatively rich; there are 70 mammalian species, 20 of which are large mammals; 90 species of resident birds, and around 100 species of reptiles.

In contrast with the Sahara and Arabian deserts, the deserts of Central Asia present fold mountains slopes cut by steep valleys with mostly ephemeral streams, fringing alluvial fans, and enclosed basins, some of which contain lakes (Caspian and Aral Seas, Lop Nor) or playas. The Turpan Depression in western China reaches 154 m below sea level. The predominant vegetation types are all formed by classic desert species of the Old World: grasses such as *Panicum*, *Poa*, and *Stipa*; sedges (*Carex*); desert bulbs such as wild tulips (*Tulipa* spp.); halophytes such as *Salicornia* and *Atriplex*; and shrubs such as *Artemisia*, *Euphorbia*, and *Caragana*. The most common vegetation found in the ecoregion is the desert sagebrush and other *Artemisia* species. The range of the flora goes from sagebrush to psammophytic (dune-adapted) plants, and includes salt-tolerant chenopod communities (Chenopodiaceae) in Afghanistan. The salt pans have almost no vegetation. In some

Figure 4.6: Palearctic deserts — the Sahara

Saharan sand dunes in Morocco.
Source: Jaime Rojo

The great shield desert of the Sahara has immense dune fields (Figure 4.6) covering 33 per cent of the total area, plains of metasediments, and sky-islands, mostly on volcanic rocks in the Tibesti, Ahaggar, and Aïr ranges. The Red Sea rift separates the Sahara from the Arabian Peninsula. The deserts of West Asia and Central Asia (Figure 4.7) are all rain-shadow deserts, mostly encircled by young fold mountains. The Sahara alone occupies 4.6 million square kilometres, or 10 per cent of the African continent. It includes an undisturbed hyper-arid central area of sand and rock, but with small areas of permanent vegetation. Vegetation tends to be much more diversified in the Western Sahara, with xerophytes and ephemeral plants in the open desert plains, and halophytes in the moister areas. Currently the

parts, the deserts of Central Asia still support small populations of rare animals like wild Bactrian camels (*Camelus ferus*) and Asian wild asses (*Equus hemionus*) that have been largely extirpated in the wild.

The Mesopotamian shrub desert is transitional between the deserts to the south and the steppes to the north. The flora includes umbrella-thorn acacia (*Acacia tortillis*), shrubby rock-rose species (*Cistus* spp.), and many dwarf shrubs. Reeds and rushes grow in the wetland areas, while poplar (*Populus euphratica*) and tamarisk (*Tamarix*) grow along river channels.

The deserts of this biogeographic realm have evolved under continual grazing pressure and most plants are adapted to grazing pressure. However, overgrazing is a persistent threat; illegal hunting is a serious threat in the South Iran Nubo-Sindian desert; egg collection and nest disturbance affect nesting migratory waterfowl. Significant portions of the Azerbaijan shrub desert and the Central Asian northern desert are farmed under irrigation, causing water and soil pollution by the use of fertilizers and pesticides. The region as a whole contains most of the world's known oil reserves and large gas reserves, increasingly exploited and with associated urban areas and infrastructure development. Groundwater resources — vital but largely non-renewable — are also under heavy exploitation pressure.

Land Degradation in the World's Deserts

TRENDS IN LAND USE AND LAND DEGRADATION IN DESERTS
All deserts have evolved under water scarcity; drought does not destabilize them. But human-induced degradation does occur — through overgrazing, clearance of woody vegetation, farming, irrigation-induced salinity, soil and water contamination by agrochemicals, groundwater exploitation, and traffic and urban-industrial-mining occupation (Figure 4.8). Apart from grazing and wood collection, these pressures tend to be restricted to relatively small areas. It is important to note, however, that these areas are generally of higher productivity, and in them the myriad needs of human societies must be met alongside those

Figure 4.8: Salinized soils

Land degradation as a result of salinization is remarkably visible in this abandoned nursery of *Washingtonia* fan-palms near the Salton Sea in the lower Colorado River valley, California. The salinization of the Salton Sea's waters as a result of concentration of agricultural drainage from the Imperial Valley reached lethal levels for plant growth and forced the abandonment of the nursery.
Source: Michael Field

of native biota. By definition, dryland farming (that is, not under irrigation) is very restricted in true deserts. There is hardly any natural accumulation of organic matter to lose, dunes are naturally mobile, and large areas have developed a stony surface lag of stones and gravel that resists further deflation and rain splash. The most important sources of dust are ephemeral river courses, the active margins of alluvial fans, and dry lake beds.

Desertification paradigms
In deserts, land degradation may manifest in a variety of ways: as changes in vegetation composition, as erosion of soils by wind and water, or as salinization and pollution associated with irrigation. The major causes of land degradation in deserts are overgrazing, wood collection and deforestation, and non-sustainable agricultural practices. The Global Assessment of Human-Induced Soil Degradation (GLASOD), based on expert opinion (UNEP 1997), estimated that 20 per cent of the world's deserts are affected by some type of land degradation.

To frame a discussion on land degradation of deserts it is important to differentiate between desertification (that is, land degradation in drylands that may be semi-arid or dry sub-humid areas) and degradation of deserts proper. As noted throughout this report, desertification is ecosystem degradation and not a process whose end product is the rich and specialized desert ecosystems that are the subject of this report. The discussion of degradation in deserts should then be placed within the context of two differing points of view: the *desertification paradigm* versus the *counter paradigm* (Safriel and Adeel 2005).

The traditional *desertification paradigm* analyzes soil and vegetation degradation as the triggers of a set of negative interactions that lead to a generalized impoverishment of the environment. According to this view, desertification takes place mainly where agriculture and intensive grazing are the major source of local livelihoods — that is, in dry-farming areas outside the boundaries of the true deserts. The loss of soil and vegetation cover leads to an associated decline in the provision of ecosystem services and a rise in poverty. Additionally, the changes in land cover and soil are also linked to increased aridity as part of a negative feedback loop, a fact that makes desertification, within this paradigm, practically irreversible, and its inevitability increases with aridity (Cleaver and Schreiber 1994).

The main tenet of the *counter paradigm* is that the interaction of direct and indirect drivers combined with the local situation, can create different outcomes. While the desertification paradigm focuses only on the negative interactions, the counter paradigm approach considers both the negative and the positive interactions, depending on how humans respond to the direct and indirect biophysical and anthropogenic drivers of change. According to the counter paradigm, the drivers, processes, and events described in the desertification paradigm do exist, but the chain of events that leads to desertification and the chain-reaction cycle of reduced ecosystem productivity and poverty are far from inevitable. The message of the counter paradigm is that the interacting direct and indirect drivers combined with the local situation can create a range of different outcomes, and that raising a general alarm based on often insufficient scientific understanding or evidence is, in the end, much less effective than identifying individual problem areas where large influxes of refugees or other complicating factors have led to unsustainable local responses. This alternative

paradigm also postulates that it is crucial to distinguish between problems originating from the natural harsh and unpredictable conditions of dryland ecosystems, such as cyclic droughts, and problems caused by unsustainable management of the environment, since the remedies will often be different (Safriel and Adeel 2005).

In general, desertification is a problem associated more with semi-arid dryland agriculture than with true arid and hyper-arid deserts. Where desertification does occur on deserts, it takes place mostly on the desert margins and less in their vast interiors. Other sections of this book discuss in detail the problems of land degradation for particular cases of desert rangelands and pastoralist societies; cultivated lands, soil erosion, desert irrigation, and oases; mining and mineral resources; urbanization, industry, and infrastructure; tourism in deserts; military testing areas and historic battlegrounds; and low impact activities in protected areas.

IMPACT OF LAND DEGRADATION

Aside from the direct effects of reduced vegetation cover, especially from overgrazing, wood collection and deforestation, human-induced land degradation does not appear to be a serious issue over the greater part of the global desert area. Salinization can be an important problem in some oases, but in true, open deserts, natural vegetation cover is nearly always sparse, there is no soil structure, and little or no soil organic matter to degrade. As will be seen in more detail in Chapter 6, most of the human-induced pressures on the desert biome tend to concentrate on the deserts' edges, that is, the transitional ecotones between deserts and non-desert regions where some productivity can be derived from the land, and in the more humid environments inside the desert biome, such as oases and desert mountains, or sky-islands. The true deserts, however, which cover the vast majority of the biome, are normally too arid and too inhospitable to be the direct target of large-scale development.

Some impact, however, does occur, and at a global scale its cumulative effect can be significant. The deterioration of vegetative cover mostly increases the amount of dust particles suspended in the atmosphere. Populations with more direct exposure to this phenomenon are more likely to develop allergies and other respiratory ailments. Sources of dust are largely restricted to severely disturbed areas, as well as some natural sources such as lake beds and ephemeral stream channels. Newly remobilized dunes become an issue for humans when they advance over settlements and infrastructure.

Military activities and off-road vehicles do cause extensive, lasting damage to the fragile desert cover. The Mesopotamian shrub desert is fascinating from the ecological and cultural perspectives. Located in the Tigris and Euphrates River valleys, it is also an important winter stopover for migrating Eurasian birds and a refuge for the endangered and sparse populations of wolves, hyenas, leopards, oryxes, gazelles and wild boars. This desert, considered a cradle of civilization, has been greatly impacted by the recent Iraq wars.

Grazing pressure on the desert, and especially on the desert margin, is the most extensive agent of land degradation. For example, the Chihuahuan Desert in Mexico is in a vulnerable conservation status because cattle grazing and browsing have damaged sensitive desert scrubs and riparian habitats. The degradation of these riparian habitats, coupled with the loss of springs as a result of aquifer depletion and the diversion of streams for irrigation, have had a great impact on wildlife that depends on water sources.

Although mining activities affect small areas directly, they have significant impacts on surrounding areas. When the mine reaches the end of its life, the site is normally abandoned and remains a mixture of deteriorated materials, mining by-products, and unproductive rubble, usually coarse, and often of extremely toxic chemical composition, which is unfavourable to colonization by plants and animals. These sites have left a legacy of polluted land and groundwater, with wide impacts through the redistribution of toxic elements by wind and flash floods, in many deserts.

Chile and Argentina have abandoned mines of copper, lead, and nitrate in the Puna that are potential sources of contamination due to inadequate rehabilitation after their closure and

the risk of chemical spillage (Romero and others 2003). The mining centres located at high altitude in the Dry Andean Puna, a mountain desert, near the source of rivers that feed irrigation systems or provide populated areas with drinking water, are particularly dangerous and require special consideration. The Rio Grande in the Quebrada of Humahuaca receives water from the Yacoraite River that drains from the eastern face of the Aguilar Mountain, which is currently being mined. The water of the Yacoraite has high levels of lead, iron, manganese, and molybdenum unsuitable for both llama grazing near the mines and the inhabitants of the Quebrada downstream (Figure 4.9). At the same time, this area has been declared by UNESCO as a World Heritage Site and is also an important tourist destination. The freshwater of the Rio Grande is used for irrigation of orchards (early season vegetables) that are commercialized in large urban centres whose population is put at risk by the metal discharges (Box 4.2).

Mining and salt extraction of sulphates, borates and others may also contribute to desert wetland pollution and consequently affect principal sources of water. Modern mining methods are very water-intensive. In addition, mining companies often excavate beneath the water-table and must pump and remove the groundwater, in the so-called practice of "dewatering." Dewatering can cause failures of springs and wells, land subsidence, and also threatens oases, wetlands and irrigation.

Figure 4.9: Mining impact in deserts

A three-dimensional view of the Mina Aguilar and the down-water Quebrada de Humahuaca in the Argentine Dry Puna.
Source: Google Earth image browser

The Laguna de Pozuelos (Pozuelos Lake), located in the Argentine Puna, has been dedicated as a 4 000 km² Biosphere Reserve within UNESCO's Man and the Biosphere Programme (MAB). Among other goals, MAB Biosphere Reserves are study areas for the scientific understanding of the complex interaction between humans and the environment.

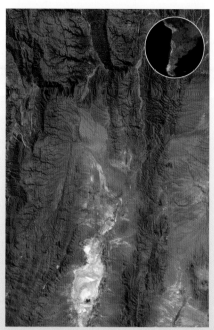

The Laguna de Pozuelos is an excellent example of an inland drainage basin with a permanent fresh or brackish water lake with fluctuating salinity. The vegetation is characterized by sparse shrubs and arid grasslands. The wet areas around the watercourses in the upper parts of many streams are covered by permanent green meadows (*vegas*) and marshlands (*ciénagos*). Thousands of the endemic Andean and James' Flamingos (*Phoenicopterus andinus* and *Phoenicopterus jamesi*) occur on the saline lakes, feeding by filtering micro-organisms with their specially adapted bills. These birds are a major attraction for tourists and bird watchers.

The abundance of species is strongly correlated with habitat variability. The threatened vicuña (*Vicugna vicugna*), listed in IUCN's Red List of Endangered Species (Baillie and others 2004) and the ostrich-like *choique* or Lesser Rhea (*Pterocnemia pennata*), graze on the steppes and the peat lands or *bofedales*. Life thrives in the freshwater lake and the briny salt flats (*salares*). Relative to today, the human population in the times of the Inca Empire was greater and denser. The main activities now are extensive grazing with camelids and sheep, and mining. Siltation, originating from water erosion in the basin due to improper management, reduces water depth (Navone and others 2004), which can have cascading effects on the ecosystem.

Landsat ETM 5 image of the Laguna de Pozuelos Biosphere Reserve and its basin in the central Dry Puna.
Source: CIATE, Facultad de Agronomía, University of Buenos Aires

The main objectives of this Biosphere Reserve are to restore degraded lands, develop ecotourism, and promote reforestation with woody species, such as *queñoa* (*Polylepis tarapacana*). In the Laguna de Pozuelos, an attempt has been made to make socio-economic development compatible with conservation, with a series of actions aimed at improving animal husbandry practices, sustainable building construction, and new methods for using natural resources.

Andean flamingos (*Phoenicopterus andinus*) in an Andean salt lake.
Source: Jean-Leo Dugast/Still Pictures

Grazing vicuñas (*Vicugna vicugna*) on a desert river bed in the Dry Puna.
Source: Rafael Introcaso

Box authors: Stella Navone, Marta Bargiela, Clara Movia, and Walter Massad

Oil spills on land and in freshwater bodies are frequent in some Palearctic deserts, and very damaging to the environment. Spilled oil affects surface resources and a wide range of subsurface organisms that are linked in a complex food chain that includes human food sources. Additionally, they can harm the environment by direct physical damage, that is, through the lethal coating of animals and plants, and by the toxicity of oil itself.

RESPONSES TO LAND DEGRADATION IN DESERTS

Inequitable access to land, human population dynamics, and poverty in developing countries are some of the most significant factors that increase overexploitation of deserts. The population of countries with large areas of deserts must often face additional challenges, both political and social, fundamentally derived from the strong competition between users of strategic water and soil resources. These situations of conflict generally result in the concentration wealth in certain sectors of society that, in time, generates territorial imbalances, lack of social equity, and, ultimately, land degradation.

Managing desert ecosystems to maintain their resilience requires an understanding of the interactions among the drivers of change, their dynamics and the thresholds beyond which undesirable changes become difficult to reverse. The key conflict lies in the struggle between human pressures and the inherent fragility of deserts, which defines the complexity of possible responses. Thus, analyses of the problems and the decisions on responses require a multi-pronged approach. It is, essentially, the art of reconciling the needs of local and global communities, and those of humans and other biota.

Responses at the global level
The first surveys of the arid regions of the world included GLASOD, conducted by the International Soil Reference Information Centre (ISRIC) under the auspices of UNEP. Much of the data generated was used in the World Atlas of Desertification published by UNEP in 1992 and 1997. One of the chapters of the recent Millennium Ecosystem Assessment (MA 2005) is devoted to drylands, and addresses, among other topics, dryland ecosystem services, conditions and trends, and drivers of change. The Land Degradation Assessment in Drylands (LADA), starting in 2006, will provide insights on the status and trends of the world's drylands. LADA is being implemented through a partnership consisting of different United Nations agencies, international agricultural research centres, farmers' associations, universities, and other civil society organizations; FAO and UNEP are jointly implementing this project. LADA will establish a standardized methodological framework to address the process of dryland degradation, increase countries' capacity to analyze and assess the causes of land degradation and areas at risk, and promote actions to control land degradation.

Many of the global conventions organized by the United Nations are responses to global degradation (Box 4.3). The Ramsar Convention, in particular, has played a strategic role on the protection of oases and other desert wetlands. However, there is no global or regional response strategy focused exclusively on true deserts. The two international conventions signed in 1992 — the Framework Convention on Climate Change and the Convention on Biological Diversity — during the United Nations Conference on Environment and Development (also known as UNCED or the Rio Earth Summit) make almost no reference to the environmental issues of deserts.

In 1994, the United Nations Convention to Combat Desertification in those Countries Experiencing Serious Drought and/or Desertification, particularly in Africa (also known as UNCCD) was adopted by the international community. This convention put a strong emphasis on the sustainable development of the arid, semi-arid and sub-humid drylands of the world, which were defined as "areas in which the ratio of annual precipitation to potential evapotranspiration falls within the range from 0.05 to 0.65." Thus, the text of the convention includes arid deserts, but excludes the hyper-arid deserts of the world as a focus of its concern, while at the same time, it includes a very large area of non-desert ecoregions where land degradation is a serious problem.

Thus, although UNCCD is not directly focused on desert environments, it does address some of the most pressing issues of land degradation that take place mostly on the deserts' edges. Additionally, UNCCD is a very important response of the international community to the threat of land degradation in drylands, and its implementation has helped in the strategic coordination and cooperation of responses at all levels — national, sub-regional, regional and international, to prevent and control land degradation, and to promote the rehabilitation of degraded areas.

Responses at regional and national levels
Apart from the international efforts that are currently being implemented, many countries are also developing their own internal policies, and making independent efforts to protect their desert environments. Although an exhaustive list would be impossible within the scope of this report, some interesting cases can be analyzed that highlight the problems of land degradation in the deserts of

The different sectors involved in the quest to stop or reverse the processes of land degradation have achieved a consensus on critical points. International agencies, national governments, non-governmental organizations, the scientific community and local associations have established a common set of guiding tools for the actions that need to be undertaken in response to land degradation. Since the 1970s, the debate on how to approach the responses to the severe problems of environmental deterioration has remained open (Abraham 2003). Among the consensus reached, and particularly for the case of deserts, the following stand out:

1. Degradation of natural resources is often associated with poverty around the world. But in many cases, degradation can be triggered by the actions of ambitious international enterprises on a major scale, like mining, oil extraction or the expansion of agricultural frontiers, which often do not generate income or improve the living conditions of local populations.

2. The use of natural resources responds to a complex articulation of social, political and economic relationships, and land degradation processes must be analyzed in the light of the historical development of local population groups, with particular emphasis on power and social conflicts.

3. An understanding of the environmental problems of land degradation in deserts requires a multidisciplinary approach. Given the complex systemic nature of land degradation processes (see Castoriadis 1997, Morin 1997, and Ciurana 2001 for a discussion on complex systems), the response must incorporate an integration of the multiple relationships among biophysical, socio-economic, political and institutional factors, including the role of population growth and inequitable land tenure among the causes of these phenomena.

4. Responses are imperative; so too are the completion of an integrated assessment (IA) of the degradation processes and the development of early warning systems. An IA is a process structured for the treatment of complex issues using the input from diverse scientific disciplines and incorporating social actors. A realistic IA requires the inclusion of those social actors likely to have influence in the decisions made at all government levels, in different administrative agencies, and in diverse interest groups.

5. The degree to which social participation is essential for land degradation abatement strategies has until recently been under-appreciated; it only emerged in strong terms in the Agenda 21 of 1992. This concept demands the inclusion of political, social, economic, and technical interests and their complex interactions in the context of land degradation. It needs to include, at the same time, an assessment of the structures of local organizations and the role played by institutions.

In conclusion, these principles are intimately related to the concepts of sustainable development and sustainable land management (SLM). They constitute the five major pillars upon which any sustainable activity must be conceived: it should be ecologically protective, socially acceptable, economically productive, economically viable, and effective in reducing the risk of degradation (Hurni 2000, GEF 2005).

Box authors: Elena Abraham and Laura Torres

different regions, and the responses on a regional basis.

In an effort to make better use of the investments in water-control structures in northern Africa, for example, a series of protection measures were implemented in watersheds in national programmes in Tunisia and Morocco. In many countries of North Africa, and also in Yemen, soil and water conservation are part of the traditional knowledge that desert societies have used for thousands of years. This knowledge has helped them adapt to aridity and drought, with sustainable land management practices that allow for soil regeneration, harvesting and conservation of water, and retention of suspended sediments in traditional terraces.

The risks of drought and water mismanagement are constant threats in deserts. To address them, there are a number of suggested technologies and practices, such as improved fallow, micro-basins, windbreaks, and earth and soil bunds. Premised on concepts of sustainability, especially in those aspects emphasizing the importance of horizontal cooperation, many of these technologies have acquired new force with more participation-oriented and integrated approaches developed since the mid-1980s. The system implemented by the local stakeholders in Gobabeb, within the Namibian desert, is a hopeful example of an organized regional response (see Box 6.4). Long-term results of research at Gobabeb focus on the variable, arid environment with a particularly high diversity of invertebrates, and contribute to understanding the basic principles underlying the functioning of arid systems (Figure 4.10).

Figure 4.10: Participatory training

The Gobabeb Training and Research Centre is located within the driest part of the Namib-Naukluft Park, and supports training and research activities from national, regional and international sources. Gobabeb is situated on the northern edge of the main Namib dune field, on the banks of the ephemeral Kuiseb River, and lies at the inland extremity of the coastal fog belt and near the western limits of scarce summer rainfall and the northern limits of occasional winter rainfall events.
Source: Mary Seely

Salinity and high sand content are major constraints in the Asian deserts. The planting of trees to help manage the spread of salt in the landscape requires large efforts. Irrigated portions of warm, arid areas in Pakistan, India, and China are major agricultural production areas but often face declining yields as a result of soil salinization.

In China, the deterioration of the plant cover in the headwaters of the Yangtze River has created major flooding problems. Massive efforts are now required to deal with the enormous problem of water erosion in the Loess Plateau, one of the most eroded regions of the world, on account of intensive agricultural practices on the steep mountain slopes. Because of the deterioration of water reserves, monitoring groundwater levels and confronting salinity problems have become essential management tools, especially in the North China plain.

In Central Asia, many of the countries of the Commonwealth of Independent States have

problems of chemical soil contamination as in Kazakhstan, salinization in the irrigation areas of the Aral Sea Basin, and soil erosion in Kyrgyzstan and Tajikistan. All have prompted organized responses centered on the identification of the processes involved and sustainable management. Unfortunately, given the political and economical crises of the region, the implementation of urgently needed responses has been slow.

Since the introduction of its National Soil Conservation Program in 1983, Australia has substantially expanded and improved its soil and water conservation technologies on private and public lands. The national and state governments continue to develop and implement sustainable land management policies, including a program of substantial reform of soil conservation, vegetation, forestry, and environmental planning law and policy. The focus of current efforts is on diversification of the commercial use of agricultural land, encouragement of conservation and remediation, strengthening natural resource management institutions, monitoring effects of changes, development of new extension and education capabilities, and controlling urban settlements on highly productive agricultural land. The national government has made a substantial financial contribution for carrying out local "government-community" conservation projects. Australia's experience has become an emblematic case of organization for controlling degradation.

In the case of Latin America, land degradation in deserts — especially salinization caused by poorly-managed irrigation systems — has been one of the main factors promoting rural emigration into urban slums, or the massive exodus of rural workers into more developed regions, looking for new alternative livelihoods. The Mexican Meseta Central, the Brazilian arid Caatinga, and the Argentine High Monte, are major sources of immigration for their countries' industrialized cities, or, in the case of Mexico, for the United States.

Concluding Remarks

Although the benefits of preventive measures may be hard to see in the short-term — due to the slow response times of the desert environment and the fact that there is as yet no comprehensive system to monitor the impacts of actions — the efforts to combat land degradation in deserts have multiplied over the years. They have become more complex with time, and are increasingly taking into consideration the natural, economic, political and cultural peculiarities of each region and country. As a result of these growing efforts, the UN General Assembly has declared 2006 as the International Year of Deserts and Desertification, in an effort to send a strong message about the global importance of conserving and protecting these unique and often fragile environments.

REFERENCES

Abraham, E.M. (2003). Desertificación: bases conceptuales y metodológicas para la planificación y gestión. Aportes a la toma de decisión. *Zonas Áridas* 7: 19–68

Archangelsky, S. (1970). *Fundamentos de Paleobotánica*. Serie Técnica y Didáctica 10. Universidad Nacional de La Plata, La Plata, Argentina

Armstrong, S. (1990). Fog, wind and heat: life in the Namib desert. *New Scientist* 127: 46–50

Baillie, J.E.M., Hilton-Taylor, C., and Stuart, S.N. (eds.) (2004). *2004 IUCN Red List of Threatened Species. A Global Species Assessment*. IUCN, Gland, Switzerland and Cambridge, United Kingdom

Castoriadis, C. (1997). *Les carrefours du labyrinthe, tome 5: Fait et à faire*. Seuil, Paris

Ciurana, E.R. (2001). *Complejidad: Elementos para una definición*. Instituto Internacional para el Pensamiento Complejo. UNESCO/Universidad del Salvador, Buenos Aires

Cleaver, K.M., and Schreiber, G.A. (1994). *Reversing the spiral: the population, agriculture, and environment nexus in sub-Saharan Africa. Directions in development.* The World Bank, Washington, D.C.

Costello, D.F. (1972). *The Desert World*. Thomas Y. Crowell Company, New York

Flores, G.E., and Roig-Juñent, S. (2002). Cladistics and biogeographic analyses of the Neotropical genus *Epipedonota* Solier (Coleoptera: Tenebrionidae), with conservation considerations. *Journal of the New York Entomological Society* 109(3–4): 309–336

GEF (2005). *Land management and its benefits — The challenge, and the rationale for sustainable management of drylands.* Document C.27/Inf.11/Rev.1, October 25. Scientific and Technical Advisory Panel, Global Environmental Facility, Washington, D.C.

Hurni, H. (2000). Assessing sustainable land management (SLM). *Agriculture, Ecosystems and Environment* 81: 83–92

Jaeger, E.C. (1957). *The North American Deserts.* Stanford University Press, Stanford

MA (2005). *Ecosystems and Human Well-being: Current State and Trends. Findings of the Condition and Trends Working Group.* The Millennium Ecosystem Assessment Series, Volume 1, Island Press, Washington, D.C.

Mittermeier, R., Myers, N., and Mittermeier, C.G. (eds.) (1999). *Hotspots: Earth's biologically richest and most threatened ecoregions*. Agrupación Sierra Madre/Conservation International/CEMEX, Mexico City and Washington D.C.

Morin, E. (1997). *Introducción al Pensamiento Complejo*. Gedisa, Madrid, Spain

Navone S., Maggi A., Bargiela M., Rienzi E., and Introcaso, R. (2004). Indicadores para el monitoreo de la desertificación en Puna y Valles Áridos. In *Teledetección Aplicada a la Problemática Ambiental Argentina* (eds. S.M. Navone, H. Rosatto and F. Vilella) pp. 97–108. Editorial Facultad de Agronomía, Buenos Aires

Olson, D.M., Dinerstein, E., Wikramanayake, E.D., Burgess, N.D., Powell, G.V.N., Underwood, E.C., D'Amico, J.A., Itoua, I., Strand, H.E., Morrison, J.C., Loucks, C.J., Allnutt, T.F., Ricketts, T.H., Kura, Y., Lamoreux, J.F., Wettengel, W.W., Hedao, P., and Kassem, K.R. (2001). Terrestrial ecoregions of the world: a new map of life on Earth. *BioScience* 51(11): 933–938 (maps available online at: http://www.worldwildlife.org/science/ecoregions/terrestrial.cfm)

Pascual, R., and Ortiz-Jaureguizar, E. (1990). Evolving climates and mammal faunas in Cenozoic South America. *Journal of Human Evolution* 19: 23–60

Romero, L., Alonso, H., Campano, P., Fanfani, L., Cidu, R., Dadea, C., Keegan, T., Thornton, I., and Farago, M. (2003). Arsenic enrichment in waters and sediments of the Rio Loa (Second Region, Chile) *Applied Geochemistry* 8: 1399–1416

Safriel, U., and Adeel, Z. (2005). Dryland Systems. In *Ecosystems and Human Well-being: Current State and Trends. Findings of the Condition and Trends Working Group* (eds. R. Hassan, R. Scholes, and N. Ash) pp. 623–662. The Millennium Ecosystem Assessment Series, Volume 1. Island Press, Washington, D.C.

Shmida, A. (1985). Biogeography of Desert Flora. In *Ecosystems of the World. Vol.12. Hot Deserts and Arid Shrublands* (eds. M. Evenari, I. Noy Meir and D. Goodall) pp. 23–75. Elsevier, Amsterdam

Stebbins, G.L. (1974). *Flowering plants: Evolution above the species level*. Harvard University Press, Cambridge, Massachusetts

UNEP (1997). *World atlas of desertification* (2nd edition). United Nations Environmental Programme, Nairobi, Kenya

5

Chapter 5: Challenges and Opportunities — Change, Development, and Conservation

Lead Author: Andrew Warren
Contributing Authors: Martin Green, Stefanie M. Herrmann,
Conrad Roedern, Uriel Safriel

Oasis, by Ali Selim
Source: Werner Forman/Art Resource, NY

This chapter explores the challenges and opportunities for deserts in the coming decades. It is about what might happen, what could happen, and what should happen in deserts. It is an emotional progression: from anxiety about the future, through excitement about its possibilities, to anger when the environment, especially its beauty, is threatened.

The Forces of Change

POPULATION

Population in deserts will change, but unevenly. Few people, mostly pastoralist nomads, live in the great spaces of the desert. Even if the high birth rates often described for some nomad groups were true, the additional numbers would be few. Much larger mining and drilling communities will also have little overall impact. They consist, and will continue to consist, disproportionately of young, short-stay men, whose numbers fluctuate with the price of minerals. In Leonora, an old mining town in Western Australia, over 60 per cent of the population is still male, a century after its foundation.

Rural groups living along the great rivers will have more impact. Scattered, long-established, smaller oases have the same demographic dynamics, but smaller numbers. The World Economic and Social Survey (UNDESA 2005) predicts a steady decline in fertility in these populations, but also overall natural growth for some years. In Egypt, for example, the number of 25–28 year-old men will grow from 3.6 million in 2005 to peak at 3.8 million in 2025, before dwindling. This will put increasing demands on resources, particularly water, and may lead to dissatisfaction with unemployment. But the production of so much potential labour has a more positive implication, because labour is desperately needed in the industrialized world: in Italy, the number of 2–28 year-old males has already peaked and will have halved by 2025; an extreme, but characteristic case. Labour-seeking industries may be attracted to these growing desert populations, but because the greatest demand for labour is in the service sector, more of the surplus will gravitate to the industrialized world, and this will boost a counter-flow of remittances. In Pakistan, remittances peaked in 1982–83, when they contributed 75 per cent to the overall balance of trade (Amjad 1986). In 2002, remittances to the developing world were

already US$67 billion, against government and bank lending of US$14 billion (Islam 2003).

Two further groups will have much greater impact. Both have grown and will grow quickly, but by immigration, not by natural increase. The smaller of these two groups is rapidly growing, and uses resources at a high per-capita rate. This group consists of retired migrants to the desert, and inhabits principally parts of the U.S. southwest (Figure 5.1). Growing numbers now also live in the United Arab Emirates (Figure 5.2). Very many more people live in cities like Lima, Cairo, Baghdad, Riyadh, Karachi, Kashgar (Kashi), Urumqi and Yarkand, all with populations of over five million. All have grown and will continue to grow quickly, and all attract many more men than women. Some depend on the production of oases (Cairo and the western Chinese cities); others are supplied from further afield. All consume large quantities of water, although all also pass on large quantities of re-usable water.

INVESTMENT AND CAPITAL

Investment (of a more formal kind than remittances) has a less certain trajectory. Inward investment was the strongest driver of change in deserts in

Figure 5.1: Palm Springs

Palm Desert

Palm Springs

0 0.75 1.5 kilometres

10 May 2000

Palm Springs, California, is where over 30 per cent of the population was over 60 years old in the year 2000. Their profligate use for water for gardens and golf courses is vividly picked out by the red ("false-colour"), which shows photosynthetically active vegetation. The un-irrigated desert is a dull blue-grey.

Source: Advanced Spaceborne Thermal Emission and Reflection Radiometer — ASTER

Figure 5.2: Burj Al Arab

The Burj Al Arab Hotel, Dubai: very high-class desert tourism.
Source: Jumeirah Hotels website, http://www.jumeirah.com/

the recent past. Most went to the extraction of oil, gas, iron, uranium, phosphates, nitrates, and copper, among other minerals. Even if no new reserves of oil are discovered, deserts contain a high percentage of global reserves, and this implies continued investment, if at a lower rate. Rising prices may maintain the income from older investments. Investment in gas is newer, and will probably increase. Iron ore contributes 40 per cent of Mauritania's export income; desert Western Australia contributed 16 per cent of the world's production of iron in 2003. Both will probably maintain their position, although iron prices fluctuate wildly. One-third of known recoverable global reserves of uranium are in Australia, but none of its desert reserves is currently mined. Namibia has about six per cent of known global recoverable reserves, and the Namibian mine is the only desert uranium mine currently in production. A global move to nuclear electricity generation would encourage the reopening of other reserves, as in Kazakhstan, Niger, and northern Chad (over which Chad and Libya went to war in 1987). North Africa (largely its deserts) holds about one-third of world reserves of phosphate.

Desert tourism, another source of investment, has grown quickly. Four million tourists visit Morocco and five million reach Tunisia each year. They contributed six per cent to Tunisian gross domestic product in 1999, and employed over 300 000 people. Desert destinations in both countries outperformed their coasts. There was a 161 per cent increase in tourism to Egypt in 2005. Dubai

claims to be the world's fastest growing tourist destination; 100 000 British people have bought homes there, and it is aiming at 15 million tourist visits a year. Baja California is booming. More gambling is said to take place in deserts than in any other global environment. The upward trends may continue in some places, although some markets must be nearing saturation. If the past is a guide, the pattern of development will be patchy.

The steady growth in short trips, as to the U.S. deserts, is less likely to falter. They attract less investment, and probably generate less income, but involve more people. An estimated 800 000 people a year, and as many as 80 000 on a single weekend, already visit the Algodones Dunes in California, many to go dune buggying. The "Burning Man", a week-long cultural festival with international draw, is held annually on the playa of the Black Rock Desert in Nevada (Figure 5.3).

Figure 5.3: The "Burning Man" ceremony, 2005

Nevada

Black Rock Desert

3 September 2005

The Burning Man ceremony is an "annual experiment in temporary community dedicated to radical self-expression and radical self-reliance" held in the Black Rock Desert about 120 miles north of Reno, Nevada.
Source: IKONOS image taken on 3 September 2005; available at http://www.spaceimaging.com/gallery/ on 12 January 2006

Large, cheap sites, which are superabundant in deserts, are a further source of investment (generally by the state). It is not only these qualities that attract this investment: it is also distance from prying eyes and air-borne attack, and freedom from planning objections. All this allows deserts to be used for purposes that would be difficult or even unimaginable in other regions; the first nuclear bomb was tested at La Jornada del Muerto in New Mexico in 1945; the

Figure 5.4: Desert launch centre in China

Jiuquan Space Launch Centre, Guansu, China

Launch Tower

6 October 2005

Chinese astronauts blasted into orbit from this desert site on 11 October 2005

Source: IKONOS image taken on 6 October 2005, available at http://www.space imaging.com/gallery/ on 10 January 2006

U.S. plans to store nuclear waste in about 600 square kilometres of Yucca Mountain, Nevada. The French tested their nuclear bombs at Reggane in the Sahara and the British at Woomera in the Australian desert (a reserved area of about 127 000 km²), where Australian nuclear waste is now stored. Later came Lop Nur in the Chinese desert (about 100 000 km²), the Kharan Desert in western Pakistan, and Pokhran in desert India. Military training takes place in the Mojave Desert, the Omani Desert and the Negev Desert in Israel. Russia, China, Japan, and the United States have space-flight installations in the desert (Figure 5.4). Low-cost space in the California deserts is used to park huge numbers of unused aeroplanes. There are prisons in many deserts; Woomera is used to hold refugees, and the European Union is said to be planning a holding station for refugees in the North African desert. These intrusions import many people into deserts, generate considerable income, and help to upgrade infrastructure, but have large environmental footprints, particularly with respect to water. In an insecure and competitive world, this kind of investment will continue, even grow.

GLOBALIZATION

Globalization may continue to be faster in deserts than elsewhere (Box 5.1), but it is also countered by rising oil prices, increasingly restricted migration, nationalism, and cultural separation. In some

---

Box 5.1: A short history of globalization in deserts

Globalization is nothing new in deserts. Linguistic uniformity, one of its key features, was established hundreds of years ago in the Old World deserts (Diamond 1997), largely because of political unification (a second key ingredient) as under Darius, Alexander, Genghis Khan, Timur and the Muslim Caliphates. Urbanization, yet another key ingredient, was a product, some believe, of the need to organize irrigation (Wittfogel 1957). The Tehuacán Valley, in central Mexico, was a cradle of New World urbanization (Plunket and Urunuela 2005). Traffic between Old World desert cities thoroughly globalized the cameleers, traders and camp followers on the Old Silk Road. They must have rolled their eyes up at globally renowned works in medicine (Al-Samarkandi), astronomy (Qadi Zada), humour (Nasr Ud Din), and poetry (Omar Khayyam), to say nothing of the silks, porcelain and musks that came from beyond the deserts. The Garamantes, who carried goods across the Sahara to the Romans, built impressive Romanesque cities deep in the Libyan Desert (Mattingly and others 2002). The Bedouin of Sinai and northern Arabia controlled yet another route between civilizations, and have been constantly re-globalized, as recently when they adopted satellite phones to play the stock market (Lancaster 1981).

Globalization reached more distant deserts, in Australia and southern Africa, in the era of railways, telegraph lines, radio networks and metalled roads. When international capital financed the exploitation of the vast reservoirs of oil beneath some deserts, some Bedouin were suddenly transformed into some of the richest, biggest-spending and best-travelled people on earth. The impacts of globalization were not all benign, for many desert people have suffered war, suppression and famine precipitated by global conflict. The recent wars in the Chadian, Angolan, Afghan and Iraqi deserts are examples (Figure). Even in desert regions far from apparent conflict, indigenous minority groups, such as the Tuareg and Sahraouis, are affected by increased militarization (Keenan 2005).

Globalization: A Russian-made rocket launcher abandoned in the desert during the proxy war between Chad and Libya (1987).
Source: Charlie Bristow

---

senses, the deserts are less globalized now than half a century ago: most were once colonial, few now are. Deserts on the tourist maps of the 1970s, as in Algeria, Sudan, Afghanistan and Iran, are now less accessible, although others like Central Asia, have re-opened. The balance of oil investment is also shifting away from deserts. The impact of globalization, where it happens, will be benign here and malign there, as before. The benign impacts include freedom to seek work, attract tourists, and benefit from medical advances and new ideas.

Inequity is a malign outcome of globalization, although it has many other roots. The outlook for equity is murky, and this should prompt concern: inequity is a powerful force for change, perhaps now expressing itself through terrorism. It is true that some remittances are invested constructively back home, as in dry parts of Kenya (Tiffen and others 1993), but remittances are unreliable, and after all, depend on inequity. Globalization counters equity in many other ways. The rich use their gains from globalization to buy off political and economic processes and remove autonomy from the poor. The overall balance between the negatives and the positives in globalization, even its strength, is debatable (Hirst and Thompson 1999).

CLIMATE CHANGE

Climate is another changing input into the equation, if a more dispassionate one. It has the potential to seriously threaten water supply, the most critical of desert resources. Climate could change at two scales. On a long time-scale, climate has and could again change without human intervention. The paleoclimatic record shows that radical change has happened within a decade (or even more quickly). The Akkadian civilization in Iraq and the Indus Valley civilization in Pakistan were brought down by sudden climatic change about 4 000 years ago (Staubwasser 2003). Only 6 000 years ago, Lake Chad, the northern basin of which is now in a literally howling desert, was a freshwater lake bigger than the present Caspian Sea (Drake and Bristow 2006).

Faster climate change is already happening and most climatologists believe that its acceleration is inevitable, whatever the cause, and almost whatever the response: it may be too late to intervene to change the trajectory of the next few decades (IPCC 2001). Temperatures, and with them evaporation (and hence aridity), will almost certainly rise further, which may or may not be compensated by increased rainfall. The deserts whose own climate is most vulnerable to change are in southern Africa. Projections for decreases in run-off in southern African rivers are of the order of 10–30 per cent (Milly and others 2005). Deserts that will most certainly suffer (and perhaps badly) are those that get their water from alpine meltwater (see Chapters 3 and 6). Climate change could adversely affect human health, both through rising temperatures or through increases in rainfall, or its variability. The virulence of plant or domestic animal pathogens may increase, or crop yield could decrease (say after drought), leading to malnutrition. Some climatic effects are well proven, as in the correlation between ENSO events and plague and hantavirus pulmonary syndrome in the U.S. Southwest; child diarrhoea in Lima; and the effects of increased ozone levels in urban areas, brought on by higher temperatures (Patz and others 2005).

ENERGY

Costs of energy, many believe, can only rise. Petrol prices rose to almost unprecedented heights in the U.S. in 2005, and despite a brief fall in early 2006, the trend is upward. The price of natural gas was a major political issue in Europe in the winter of 2005–6, and most commentators expect natural gas prices to rise further. The costs of aviation fuel are threatening the financial viability of many airline companies. Although some deserts command large reserves of oil and gas, most do not. Dearer energy will affect them all, but in different measure. In deserts, where the relation between the price of energy and that of water is close, more expensive energy will restrain many development schemes. The costs of solar and wind energy may be lower in deserts than in some other places, but even so, they do not yet compete well with fossil fuels, except in a few sites, and their cost-trajectory is uncertain, especially if cheap fossil energy is not available to build the necessary facilities. The costs of travel seem bound to rise, so that tourism may suffer (among many other things), although rising travel costs are claimed not yet to have had an effect in the U.S.

RESTORATION

The costs of cleaning up the mistakes of the past have been rising and are likely to continue to do so. There have been many of these mistakes, some of them discussed elsewhere in this chapter. The reclamation of salinized land, and the revival of economies that once depended on that land, already consume large sums, and could consume much more. By far the best-known case has been the Aral Sea basin (Figure 5.5), which will take decades to restore (if it is ever achieved). The existing recovery programme will only save one basin of the former sea, and reduce only a proportion of the dust that damages health (Kingsford and others 2005).

Figure 5.5: The vanishing Aral Sea

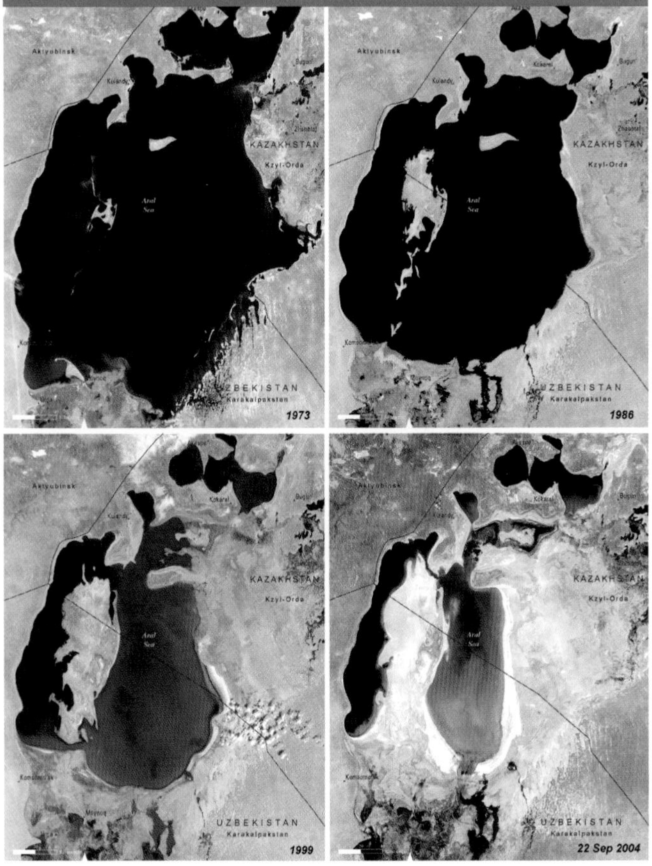

The Aral Sea began to contract after a large proportion of the water in the two main feeder rivers was diverted to irrigated cotton in the 1960s and 1970s. By 1987, its level had fallen 14 m; its salt concentration had doubled from about 10 grams per litre in 1961 to 40 grams per litre in 1994; 20 of the 24 native fish species had been lost and there was a virtual end of commercial fishery; dust storms became toxic with salts and agricultural pesticides; 97 per cent of women in the surrounding area are now anaemic; life expectancy is significantly lower than in surrounding areas. The rescue effort includes the re-engineering of the Syr Darya River delta in the north, which will retain water in the northern basin, but desiccate the South Aral Sea, perhaps within 15 years.

Source: Kingsford and others 2005; image: UNEP/GRID, Sioux Falls

The Aral is not unique. "Salinization and waterlogging have affected 8.5 million ha or 64 per cent of the total arable land in Iraq; [...] 20–30 per cent of irrigated land has been abandoned due to salinization" (UNEP 2002). In the Tarim River basin of China, more than 12 000 square kilometres of land was salinized between the 1960s and 1990s (Feng and others 2005). In both countries, degradation is expanding. Because salinization slowly and incrementally affects yield well before it precipitates complete failure, the problem may take years to become apparent to anyone but the local farmers. Collapse of the whole may then be sudden. This is what is said to have happened in Iraq, 3 000 years ago, where the legacy of salinized soils is almost as great today as it was then (Jacobsen and Adams 1958). Dealing with accumulated salt may take centuries, if not millennia.

The prospect of wars over water (Bulloch 1993) has not materialized. Countries that compete for water, like India and Pakistan, have kept strictly to their agreement about sharing water (in their case, the Indus Waters Treaty), even if they have resorted to war over other things. Perhaps water is too important to fight over (Alam 2002). This is not to say that water is not a source of local conflict, as it has been recently in Cochabamba in Bolivia, and continues to be in Israel and Palestine. Reclamation may not, then, be violent, but it will not be cheap. Communities and economies will need to be relocated; national and regional economies have to be adjusted — some of them for the better in the long-term, if we believe Reisner (1986); more will have to be spent on supplying water, and on international treaty obligations. Dust, as from the Aral Sea and the Owens Lake in California, has to be controlled. Sedimenting reservoirs have to be managed more carefully and on a longer-term basis.

Collapse

Accounts of the collapse of the Iraqi and Indus Valley (above) and other desert collapses, such as that of the irrigation system in Chaco Canyon in Arizona in the 12th century (Diamond 2005), show that collapse is a common hazard in deserts. This is largely because desert economies depend so utterly on water. In ancient water-supply systems, withdrawal or degradation usually brought collapse. Vulnerability may now be reduced, but it is not eliminated, as the Aral crisis shows.

CULTURAL CHANGE

In a history of the Grand Canyon, Stephen Pyne (1998) traced a progression of different aesthetics: indifference among the early Spanish explorers; revulsion at a wasteland among the first potential settlers; incomprehension of a huge depression, when landscape taste had discovered mountains; wonder at its geological history, following the discoveries of John Wesley Powell; the American sublime. Each desert has had a similar history of changing tastes, and each will continue to experience this kind of fundamental change.

The Challenges and Opportunities of Development

This section is about what could happen. Technological and organizational changes could bring large gains, but a lesson should be learnt from the short review of dreams about deserts in the past: many promises come to nothing (Box 5.2).

WATER

Water has always been and will continue to be crucial to the development of deserts. But opinions about water development have changed. Fifty years ago there was immense faith in engineers and in state investment to support them (in the western, socialist, oil-rich, and Third World countries alike). But this faith closed planners' ears to warnings about the long-term impacts and real costs of water projects (Reisner 1986). Today, this legacy considerably constrains opportunities in three ways. First, the need to clean up the consequences, as discussed; second, the need to extricate policy from the past and to construct better policy for the future; and third, the depleted and degraded state of whatever water remains.

Building more dams and drilling for more groundwater still tempt the policy-maker, and in many cases the temptation is undeniable. The ongoing debate in Pakistan about a proposal for

a dam at Kalabagh on the Indus system illustrates the problems of this course. The water in the Indus and its tributaries is already thoroughly utilized; demand for water is rising with population and modest increases in prosperity; thus the distribution of water is increasingly controversial. The huge dams that were built with foreign aid after the signing of the Indus Waters Treaty in 1960, as at Tarbela (on the Indus itself) and Mangla (on the Jhelum), are filling with sediment (Tate and Farquharson 2000). The first dams took the best sites, leaving sites that are less than optimal. Large, impounded reservoirs lose large amounts of water by evaporation. Dam-building requires huge investment, with questionable long-term returns, and no guarantee that the water they save will not lead, as the water from earlier dams did, to salinization. Climate change will probably decrease the flow of the snow- and ice-fed rivers that are Pakistan's main sources of water.

A high proportion of desert river water is already used. The Colorado River in the USA, the Nile and the Mesopotamian rivers are now nearly completely utilized. Siltation of reservoirs happens worldwide; in the next few decades, sediment may clog the outlets of the Glen Canyon dam (inaugurated only in 1963) on the Colorado River in the U.S.; some NGOs are already calling for its decommissioning. Almost all the sediment in the Colorado and the Nile is retained; the figure is 60–80 per cent in the Tigris-Euphrates system (Vörösmarty and others 2003). The time between the completion of large dams and the point where siltation threatens to close them is running at about 30 to 40 years. If so, many more will soon be threatened.

Returns on investment from water schemes are debatable. In the Central Valley of California, state subsidies on water, in one case, and at one time, amounted to about $217 per acre per year, for land that yielded crops with a value of only $290. The crop was cotton, a highly water-consumptive crop, in national surplus and grown more cheaply in other parts of the USA, let alone elsewhere in the world (Reisner 1986). Of course, dams yield more than water, particularly hydroelectric power and recreation, and these must always feature in cost-benefit analyses, but all of these benefits are threatened by siltation, an almost inevitable

Deserts, to those who do not know them, are blank slates. To the desert fathers of the early Christian church, the blank slate was to be filled with God (Figure 1). Imperialists saw something else on the blank slate: somewhere to plot the straight frontiers of their new political entities. Other Europeans and North Americans — scientists and others — had nightmares about the environmental decay they thought they saw in deserts, and used the desert as an object lesson in environmental management (for example, Percy Shelly's "Ozymandias", published in 1818, Peter Kropotkin 1914; Hedin 1931, Lowdermilk 1943).

Dreams then turned environmental and technological. The dream of deserts reclaimed by planting trees, prompted President Roosevelt to propose a 260 × 3 000 km greenbelt across the Great Plains (Zon 1935). Edward Stebbing (1938) dreamt of one across Africa. Much later, the Japanese dreamt of a greenbelt in the northern Sahel (Rognon 1991). More technological dreamers proposed to paint kilometre-scale squares of desert black: dark surfaces would be hotter than light ones; the heat would lift the air; this would encourage rain (Glantz 1977). This idea was less successful than the notion of cloud seeding to bolster rainfall, but it too came to little (Silverman 2003).

The dream of diverting water to reclaim the deserts came true in many places, yet it turned to a nightmare in some, like the Aral Sea. More elaborate dreams were never realized: the Amazon taken in a pipe beneath the Atlantic to West Africa; the Congo pumped over the hill towards Lake Chad; the Zambezi to the Kalahari; the Ob and the Yenisey from Siberia to the deserts of Central Asia (a project that has again been dusted off); the Alaskan rivers to the southwestern United States (another dream that cannot be shaken off). In another dream, seawater was channelled into desert depressions, to test whether this would enhance rainfall. The Chott Djerid in Tunisia, an early target, proved to be above sea level, but the Qattara depression in Egypt was indeed below sea level, and is still the object of these dreams. This kind of dream meets three kinds of waking reality: vast investment; poor support in science; and a history of disaster among earlier dreamers (Glantz 1977; Figure 2).

Figure 1. The search for God in the desert. The Santa Katarina Monastery, at the foot of Mount Sinai in Egypt.
Source: Andrew Warren

Figure 2. Desert dreams: Like ghost-witnesses of a glorious past, only a few dead palms remain from an old deserted golf course near the coast of the Salton Sea, in California. The place was abandoned when the waters of the Salton became saline from the accumulation of agricultural drainage.
Source: Michael Field

Dreamers of water beneath the sands were spurred on by the discovery of the Great Artesian Basin in Australia, and of deep groundwaters in the Algerian Sahara in the late nineteenth century. In Libya the dream may have been realized. Even pessimists give the Great Man-Made River in Libya, the first phase of which was completed in 1991, a century of function; optimists claim five centuries. But its aquifer, like almost all desert aquifers, has only a limited life. The deeper water is thought to be 200 000 to 1 200 000 years old, and the water held at depths of less than 600 m below the surface is about 160 000 years old. The water in the Great Artesian Basin is 225 000 to 400 000 years old. Most desert aquifers are smaller, and most are rapidly retreating under heavy pumping. The main aquifer in Shiyang in northwestern China also contains water that is largely "fossil", and is being depleted rapidly (Ma and others 2005). Moreover, many desert aquifers, like the Great Artesian Basin, contain highly mineralized water, quite unsuitable for irrigation. As Saudi wells are pumped ever lower, more and more mineralised water is brought up. Depleting aquifers in coastal areas allows in seawater, which has now penetrated 20 km inland in Libyan coastal aquifers (Allan 2005).

consequence of dam-building. There are many other examples in which the underpricing of water, or the non-collection of fees for water (for whatever reason) has encouraged profligate use, which in turn may accelerate salinization and waterlogging (Ray and Williams 1999). Many of these stories would have applied to Australian water management in the past,

but there has now been radical reform aimed at the more effective and less environmentally-damaging use of water (Turral and others 2005).

Groundwater, with some exceptions (Box 5.2), also has problems. Saudi Arabia's trajectory on groundwater illustrates some of these. The city of

Riyadh is close to a large supply of groundwater, yet the government chose to sell this water cheaply for agriculture (to promote food security), and to supply the people of Riyadh with desalinated water pumped up from the ROPME Sea Area, at much greater cost. As in the history of desert water development in the USA, the cheapness of the water allowed Saudi farmers to grow highly water-consumptive crops and even livestock (Figure 5.6; Allan 2005).

Water supply can be improved only by combining technology and management. Some technologies (Box 5.3) may make a great impact locally (as for new hotels or isolated settlements), and could play a greater part in this role. Some are only feasible at large expense and high consumption of energy. A small number, like the desalinisation of brackish water, are both cheap and widely applicable. Better policy will depend on learning the lessons of the many twentieth-century water policies that

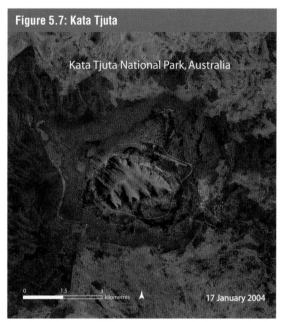

Figure 5.7: Kata Tjuta

Kata Tjuta National Park, Australia

0 1.5 3 kilometres   17 January 2004

Kata Tjuta National Park, 450 km southwest of Alice Springs, Australia. The main attraction, Uluru, is the vast quartzite rock in the centre of the image. It is an Aboriginal sacred site and Australia's most famous natural landmark.
Source: IKONOS satellite image available at http://www.spaceimaging.com/gallery/ on 2 December 2005

Figure 5.6: Non-renewable use of water

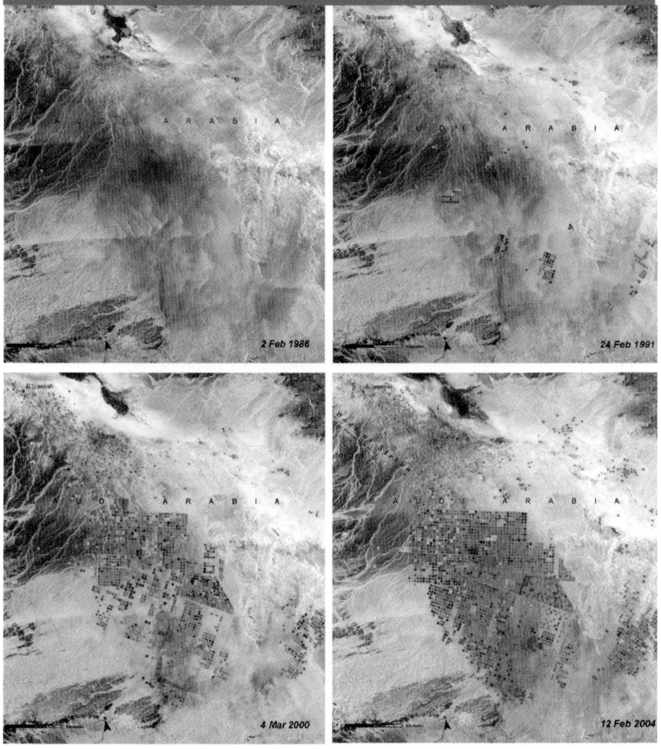

2 Feb 1986   24 Feb 1991

4 Mar 2000   12 Feb 2004

The expansion of centre-pivot irrigation systems in Saudi Arabia, where diminishing supplies of groundwater are being used to grow very water-consumptive crops, while Riyadh is supplied by desalinated water pumped up from the ROPME Sea Area. Some of the water in the aquifer is as much as 20 000 years old and therefore non-renewable.
Source: UNEP/GRID, Sioux Falls

squandered rather than conserved water. Given the escalating water crisis in many deserts, caused not least by the old policies, better policy is urgent.

TOURISM

Tourism (Figure 5.7) is another opportunity for development, but investment in tourism is risky. The risks are illustrated by the recent history of tourism in Chad and Niger. They have uniquely attractive deserts, but civil war throttled tourism in both in the 1980s and early 1990s, and the rebound, if any, has been from a low base (Hosni 1999). Chadian tourism is under retreat from unrest again in early 2006. And even if there has been growth, it can falter: visits to Namibia fell appreciably in 2004, perhaps because of recession in parts of Europe, perhaps because of reports of rising crime in Namibia itself. Tourist numbers fell significantly in Mexico, Chile and Peru after 11 September 2001. Distance is not, yet, an obstacle: large numbers of Europeans, North Americans, Japanese, and now Chinese and Indian tourists still visit distant deserts (like the Namib, the Atacama, Uluru, Oman and Dubai), but energy prices threaten all these flows. Terrorism and recession are other, inevitable threats.

Water technology is continuously improving: there have been advances in the systems for building dams and pipelines in saline conditions (which are common in deserts), as demonstrated recently in the construction of the Great Man-Made River in Libya (Hurley and Blake 2002), for lining canals, greater water-use efficiency and so on. Irrigation efficiency has been increased with drip irrigation and microsprinklers, which achieve water use efficiencies of 95 per cent, compared to efficiencies of 60 per cent or less in flood irrigation (Vickers, cited in Gleick 2001). Desalinisation of water is now mostly by reverse osmosis, but to desalinate seawater in this way costs about eight times that of getting water from conventional supplies (in wet climates). Even if costs fell, they would soon meet the rising costs of energy, of which desalination is very consumptive. In December 1995, the 11 066 desalinization plants in the world had the potential to produce 7.4 billion cubic metres per year, a mere 0.2 per cent of world water use. However, the physics of reverse osmosis mean that costs fall rapidly as the salinity of the input decreases. To desalinate brackish water may need only 0.02–0.10 US cents per cubic metre of treated water; this makes the process much more competitive (Gleick 1998). Treating and reusing wastewater is already practiced in some regions and has great potential. In Israel, 70 per cent of municipal wastewater is treated and reused for irrigation. In California, golf courses and crops are now irrigated, and aquifers are recharged, with recycled wastewater. Windhoek uses recycled wastewater to supplement municipal water supply (Martindale, cited in Gleick 2001).

"Alternative" methods of enhancing the supply of water include fog harvesting in cool-current coastal deserts (Figure). In Cape Province in South Africa, most fog occurs below the 200 m contour. The highest rates of water collection from this fog are over 2 litres per square metre of collecting surface per day (Olivier 2002). Wilder proposals include the towing of icebergs or large bags of freshwater to desert coasts, filling oil tankers with water for their return journeys from the wet world, and the direct use of seawater for irrigation (D'Amico and others 2004). Given their enormous dependency on energy, these systems are unlikely to water much of the desert.

Fog and fog collectors in Chile. Fog (*camanchaca*) is common on the coast of the Atacama Desert, between, 8°S and 32°S, as in some other coastal deserts. The experiment with the fog collectors was to see if they could produce enough water for livestock, afforestation, water supply and wildlife.
Source: Guido Soto and Waldo Canto

Tourism development has other dangers that should be recognised explicitly in policy, such as corruption (of all sorts and at all levels), competition for resources like water, damaged beauty and biological value, temptation for street and organized-crime, flagrant inequity, and litter. Against all that, ecotourism, to which the deserts have much to offer, is said to be the fastest-growing sector of the tourist market, although there are concerns that the label is used to cover activities that damage biodiversity, such as off-road motoring. More carefully defined, monitored and marketed, it has the potential to enhance nature conservation, and to contribute to local incomes (Goodwin 1996).

THE USE OF DESERT SPACE

Space-consuming installations may still bring investment, although decisions are usually taken by national governments, whose fixations with national prestige or perceived vulnerability usually blind them to the interests of desert people or environments. Many large desert voids remain, far from habitation, and contain nothing to interest tourists. Many are remarkably resilient, especially those occupied by sand dunes. Desert dwellers and tourists will undoubtedly raise objections, nonetheless, on the grounds of competition for water, safety, pollution and aesthetics, if not of peace.

Wind and solar energy installations can also make use of cheap space, large inputs of solar energy, some windy sites and the absence of objectors (Box 5.4). Small solar cells, for domestic use and telecommunications, are now quite common in deserts, but only a very small fraction of the touted potential has been harnessed. The best-case scenario would have the deserts becoming the globe's principal suppliers of energy — forgetting for the moment the environmental costs and the destruction of beauty. Against this will be the length of transmission lines and competition from other renewable sources nearer to large centres of population. At the least, it can be confidently predicted that the adoption rate of small solar and wind energy devices will accelerate, especially if the obstacles are overcome: high capital costs; expensive, short-life batteries; inadequate facilities for repair and maintenance.

Agriculture and horticulture are already profitable in many deserts, as in Israel and Tunisia, and have great further potential. If there is water, or better, re-used water, and if it is used effectively, as in greenhouses, or by drip irrigation, the intense solar radiation and seasonal patterns of low-latitude

Box 5.4: Solar and wind energy

Renewables (including solar and wind energy) provided only 0.5 per cent of world energy consumption in 2004 (Smalley 2005), but with the decline in the production of fossil fuel, renewables will have to be more productive. Their cost has been steadily decreasing and solar energy might become competitive with gas in about 2025 (GAC 2005). Renewables might supply one-third to one-half of global energy by 2050 (Shell International 2001). Deserts in general have the highest levels of solar input in the terrestrial world (cloudiness and dustiness reduce radiation in some). They also have cheap and plentiful space, far from people. Figure 1 shows the area that would be required if photovoltaic cells (converting sunlight directly into electricity) were the only suppliers to a range of markets.

Solar energy is very unlikely to be deployed on sites as large as in the figure, but some moderately large projects are in progress: Los Angeles may soon get electricity from a 1 800 ha site near Victorville in the Mojave Desert. 50 MW capacity might be installed by 2008. A task force of the International Energy Agency believes the optimum size to be yet smaller (about 5 MW; Kurokawa 2003). The northern Sahara is a good place to start pilot projects of this size, because it is already connected to the European electricity network (Cova and Kerviskadic 2005).

The greatest attraction of solar energy is that it can be captured in small installations, close to where it will be used, and free of centralized control (and malfunction). In these situations costs are already competitive: an estimate for the installation of a solar hybrid system (including subsidies for energy efficiency measures) at Tsumkwe in north-eastern Namibia is US$2.1 million, whereas merely to install a power line to the site would cost US$6.5 million (Conrad Roedern, personal communication). The major element in the operating cost of these small systems is the replacement of batteries, but they can be expected to operate for 25 years before replacement. An attractive use of solar power is for small water pumps, but once installed, maintenance of such vital services becomes critical (Hamidat and others 2003).

For wind energy, deserts have little advantage over other sites (particularly coasts), except as regards cheap space, and more

Figure 1. The Sahara with the sizes of "solar farms" necessary to supply the whole world (largest square 800 × 800 km), Europe (320 × 320 km) and Germany (180 × 180 km, smallest square).
Source: Ludwig Bölkow Foundation, with thanks also to Conrad Roedern

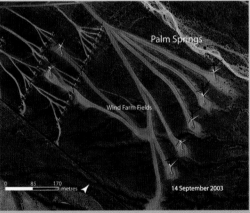

Figure 2. Wind Farm in the San Gorgonio Pass, California.
Source: Space Imaging's IKONOS satellite, http://www.spaceimaging.com/gallery/ioweek/archive/05-02-07/index.htm, 14 September 2003

freedom from aesthetic objections. As with solar energy, lengthy power connections are a disadvantage. Desert winds are strongest where the wind is funnelled through mountain passes, such as the Tehachapi Pass in the Californian desert, where there is already a large wind farm (5 000 turbines, generating 1.3 billion kilowatt-hours of electricity per year, Figure 2); there are others in the San Gorgonio and Altamount Passes nearby, all quite close to a large market in southern California.

Aquaculture is a fast-growing food production system, and most of it is happening in developing countries (FAO 2000). Surprisingly, aquaculture thrives in deserts, like the Negev and Arizona (Kolkovsky and others 2003). Deserts offer many advantages. The amount of water invested in harvested fish is less than in the production of a vegetable crop. In a vegetable crop most of the water added goes in transpiration and evaporation; fish do not transpire and do not require light for the production process, so that their containers can be covered to minimize evaporation. Many aquaculture crops are far more tolerant of, even thrive in, water that is too saline for most cultivated crops, and many geothermal waters are both warm enough and contain the right salts for fish. The dangers in aquaculture, especially of nutrient-rich effluents and pesticides, are easily controlled.

Aquaculture produces many valuable products: edible fish, ornamental fish, edible shrimps (and other crustaceans), and algae for nutritional, pharmaceutical, or industrial use. Micro-algae, which photosynthesize, fare even better than fish and

Algal culture in transparent tubes at Kibbutz Ketura in the hyper-arid southern Negev Desert, Israel. The organism is *Haematococcus*. The pigment it produces is Astaxantin.
Source: Rachel Guy

crustaceans in the drylands (Figure). They thrive in transparent tubes, which allow them to make use of the desert light. They rapidly reproduce and provide nutrition, vitamins, antioxidants and pigments. Micro-algal aquaculture can be integrated with fish aquaculture, because the larval stages of many fish species consume micro-algae. They can even substitute for fishmeal in neighbouring aquaculture systems.

The expertise and investment to start an aquaculture venture are not great. The system evolved in developing dryland countries, with low levels of investment. In Egypt, aquaculture is thousands of years old, and in Chad, people have cultivated the micro-alga *Spirulina* for their own consumption for centuries.

deserts allow them to produce when higher latitude agriculture is not productive. Aquaculture can also flourish (Box 5.5.). Another resource is wild (and newly domesticated) desert plants (Box 5.6).

Conservation and Sustainable Use

This section discusses what should happen in future desert development. It examines the conservation of soil, biodiversity, and beauty, but another general warning is necessary for those who think such an ideal is easy to achieve. Deserts are one of the last major scientific frontiers, and it can be argued that more is spent on researching the polar environment than on the more biologically-rich global desert. We do not yet know enough about the environment to be confident in prescriptions for conservation.

EROSION

Most erosion in deserts is beyond feasible control. Some of the huge areas of bare rock, steep bouldery slopes, sand dunes and pebble-covered plains barely change from year to year; others

Figure 5.8: Millet field in arid Sudan

Jebel Gehaniya, 10 km east of Kagmar in arid North Kordofan, Sudan. The arrows point to a tree that occurs in both pictures. The landscape after a very dry year (1984 in the upper image) is contrasted with the remains of a crop from a much wetter year (1989 in the lower image).
Source: Lennart Olsson

Wild desert plants are useful for many more purposes, and for many of which are less destructive than grazing, browsing or firewood. A re-examination of some ancient systems of exploiting them has led to some commercial successes, and the deserts of Mexico are a great case study for this. Wild chilies (*Capsicum* spp.), and wild oregano (*Lippia* spp.) are gathered for food and sold in Mexican and U.S. markets; wild yuccas and agaves for fibre; and plants like *Euphorbia antisyphillitica* in the Chihuahuan Desert, for waxes. Mesquite (*Prosopis* spp.) can produce 1 000 kg/ha of edible and nutritious pods (Felger and Nabhan 1980). Cacti, columnar cacti in particular, have edible, tasty, and nutritious fruits with high-quality sugars. The fruits of *Opuntia* spp., the prickly pear, are widely consumed from wild and cultivated plants. The species also yields edible cladodes (stem-pads), which are harvested in Mexico, and some are exported. The leaves of a desert aromatic, the Baja Californian damiana (*Turnera diffusa*) are now actively and sustainably harvested to make a sweet liqueur that has a growing demand. Perhaps the biggest success story is mezcal, a liquor that is distilled from pit-roasted wild agaves. Some mezcal is the drink of connoisseurs who have graduated from tequila (from the boiled stem of another, cultivated agave). There is potential for much more.

Another valuable trait in desert plants is efficient water use. The efficient tepary bean (*Phaseolus acutifolius*) has been passed over in favour of less water-efficient crops (such as the commercial cultivars of *Phaseolus vulgaris*), but the tepary has enormous potential. Indigenous mechanisms like these have the potential to produce crops for much more water-efficient agriculture. There is a long list of possible candidates for new water-efficient crops (Felger and Nabhan 1980).

Searching for distinctive germplasm takes this quest into another dimension. Germplasm that encodes adaptations to desert conditions is of particular interest to plant breeders (to produce crop varieties better adapted to the desert), and to pharmacologists, who search for biologically-active secondary compounds. The search can be helped by studies in ethnobotany, but, promising as this search may be, little seems yet to have been found — perhaps because the pharmaceutical industry is secretive in their searches.

In the Mezquital Valley of the Mexican Central Plateau, a woman combs the fibre of the lechuguilla (*Agave lecheguilla*), a desert plant, on the spines of a barrel cactus (*Ferocactus*), known locally as *biznaga*. The fibers have been previously tinted with a crimson dye derived from cochineal (*Dactylopius coccus*), a scale insect that thrives in prickly-pear cacti of the genus *Opuntia*.
Source: Fulvio Eccardi

In many desert countries, the marketing of wild products is still a major economic activity. A colourful mixture of natural herbs and manufactured products is displayed for sale in a souk, Marrakesh.
Source: Peter Tarr

erode, sometimes quickly, but the processes are seldom accelerated by human agency. In some agricultural systems, like run-off agriculture, erosion is actually welcomed (Evenari and others 1971). Rain-fed agriculture is only possible on the desert margins, and here too there is an equivocal message about erosion. Farmers in the northern Kordofan province of Sudan can produce crops from sandy soils in arid areas, but only in wet years (Figure 5.8). The sandy soils are easy to cultivate with hoes (usually the only ground-preparing tool available) and have good water-holding and water-yielding characteristics. The sandy fields may suffer wind erosion in dry periods, but many of the sands are deep enough to withstand millennia of erosion before they become unproductive. There are many such soils in arid areas. Few farmers can be convinced that erosion of these soils should be controlled, and anyway have inadequate labour to invest in it (Warren and others 2001).

Erosion, of course, is what produces the sediment that silts reservoirs. Some of it comes from high, tectonically active mountains, like the Himalayas, as in the case of the Tarbela and Mangla dams. Here, the natural rate of erosion on slopes and in river channels vastly exceeds the erosion caused by people; conserving agricultural soils would make little impact (Ives and Messerli 1989). In the U.S. Southwest, desert areas feed some of the sediment that is filling reservoirs like Lake Powell (behind the Glen Canyon Dam), and here too it appears that soil conservation would have little impact. Fifty years ago the consensus of scientific opinion was that overgrazing had exacerbated erosion in the Southwest, and if this were true, it might have been controllable. Some 30 years ago the consensus shifted to the belief that most erosion occurred in years with intense summer rains. These wet years have now been linked to El Niño cycles (Hunt and Wu 2004). Moreover, it appears that erosion rates in parts of Arizona are unaffected by vegetation cover in the short-term (Ritchie and others 2005). A more effective way of managing sediment in reservoirs is to build small sediment-holding dams upstream of the main dams (Catella and others 2005). Small dams beneath the water could hold back siltation from the outlets of the main dams (as proposed for the Tarbela and Glen Canyon dams). Although effective in the short-term, neither of these systems will give more than temporary relief.

Dust is the most significant natural output from the deserts (see Chapter 3 and Uno and others 2005). Should/could it be controlled? Control would be barely feasible for the quite natural processes that create about 90 per cent of dust in areas like northern Chad (Figure 5.9) or western China (Zhang and others 2003). Moreover, control, in the unlikely event that it succeeded, would interfere with the role of dust as a global fertilizer, as in the agro-ecosystems and forests of West Africa, the forests of the northeastern Amazon basin, the forests of Hawaii, and, most critically, the oceans, where the supply of iron-rich desert dust regulates biological productivity and may thus help to control global CO_2 (Dutkiewicz and others 2006).

Dust from agricultural operations can be a severe and expensive nuisance, but seldom in deserts.

Figure 5.9: Dust storm over Lake Chad

A large dust storm in April 2004 streaming from the northern basin of the now-dry Lake Mega-Chad, the dustiest place on Earth (Giles 2005). If irrigated plantations were chosen as a method to control the dust (the most common method), water would have to be brought 400 km from the nearest available perennial supply in the present Lake Chad (at the bottom of the image) to the origin of this dust storm. Other dust-generating deserts are also far from water supplies off sufficient size to control them.

Source: http://earthobservatory.nasa.gov/

In deserts, dust is a real pollutant downwind of desiccated lakes, such as the Aral Sea and Owens Lake in California, and here control is feasible and desirable. It is a taller order to ask for the control of military manoeuvres in the desert, like those that created large clouds of dust in the desert war in North Africa in the 1940s and the two Gulf Wars.

MANAGING IRRIGATION

Irrigated soils are a much higher priority for conservation. In most large irrigation schemes, water from rivers, wells or qanats is fed by gravity to low-lying sites. The soils in these sites are vulnerable to salinization, because the added water raises the water-table within them. When the soil water nears the surface, some is drawn up further by capillarity and the salts it carries are concentrated by evaporation on the surface, where they reduce yields, ultimately to very low levels (Figure 5.10). Waterlogging also damages crops. In large schemes, where settlers are unaccustomed to irrigation, or where planners have ignored warnings about salinization, large areas have gone out of production.

In low-intensity irrigation, as in Iraq in the 3rd millennium BCE, salinity was kept at bay by fallowing the land. During the fallow, weeds and natural drainage drew down the water-table. In

Figure 5.10: Salinization of desert soils

A Sindhi cowherd minding his cattle as they graze on a former rice field, which now has strongly salinised and infertile soil (white salt is at the surface). What was a productive field can now only support salt-tolerant wild species, like tamarisk.

Source: Andrew Warren

second millennium BCE Iraq, a centralized state replaced the ancient system with irrigation from a large canal. Salinity accumulated and steadily reduced yields (Jacobsen and Adams 1958). In intensive modern systems, where water is added to three or more crops a year, the best strategy to manage salinization and waterlogging is to maintain drainage through the soil, and to add enough water to carry away the salts. The bigger the scheme, the more expensive are the necessary drainage pipes, ditches, or tube-wells for lowering the water-table, and the canals to take salty water to the sea or to reservoirs where it can evaporate. Complementary strategies include using brackish water to irrigate salt-tolerant crops, and blending saline with fresh water, but there are dangers in both.

RIVERS, WETLANDS AND LAKES
Rivers

Wet desert environments are biologically the richest places in the desert. The richest of all are the perennial rivers. Ephemeral rivers have value, and although they profoundly affect the lives of the communities and wildlife that live near them (Jacobson and others 1995), their value and vulnerability are small compared to those of the perennial rivers, the good management and conservation of which is of utmost importance.

The most difficult issue in conserving the biological value of perennial rivers is the assurance of flow. Some, like the Tarim River, were desiccated by irrigation decades ago; most of the others have been progressively depleted. But the mere suggestion that water should be left for wildlife is risible to most managers of irrigation schemes. The second, closely linked issue is the quality of the water, and in this, the interests of conservationists and farmers come closer together. The main contaminant is "return flow": water returning to the river from irrigated fields. Return flow is always more saline than the water taken from the river in the first place. Saline return flow increases problems for downstream farmers, as when it flows from the Punjab and Sindh in Pakistan. In the Colorado River in the U.S., salinity was a minor problem in the 1960s; by the 1980s, nearly 40 per cent of its salinity came from return flow (Law and Hornsby 1982). The issue is international on the Colorado River (and on the Rio Grande that flows between Texas and Mexico), because the U.S. and Mexico agreed, in 1974, that the salinity of cross-boundary rivers should be kept below a threshold, but the salinity of water that flows to Mexico is, even so, above that threshold. The U.S. Congress, long ago, allocated US$1 billion to the problem (Reisner 1986); a desalination plant at Yuma (cost US$256 million) opened in 1992, but closed eight months later, partly because of high running costs. Hopes of a reopening are now being raised. In China, the maximum annual salinity in the upper reaches of the Tarim River rose from 1.3 grams per litre in 1960, to 4.0 in 1981–1984, and to 7.8 in 1998, as it was increasingly supplemented by return flow (Feng and others 2005). Salinity damages the ecology of perennial rivers and their floodplains. In the southwestern United States, the salt cedar, an alien, salt-tolerant tree, has invaded hundreds of thousands of hectares of alluvial plains, altering their avian and invertebrate ecology. Attempts at control have met with mixed results (Shafroth and others 2005). Return flow may also carry residues of agricultural chemicals and toxic trace elements.

Dams themselves severely interfere with the ecology of the perennial rivers. They deprive rivers of sediment, and thus depleted, most rivers excavate their channels, and this isolates and desiccates their former floodplains. When silt-deprivation is added to the replacement of cyclical floods and low flows by the more constant flow from a dam, other problems appear. After the closure of the Glen Canyon dam on the Colorado River, riffles (where the river flows shallowly over stones) expanded at the expense of pools and this favoured fish that spawned in gravel over those that did not (Magirl and others 2005). The combined effect also discourages lateral migration of the channel (as when meanders move), and this interferes with the ecology of the early-successional native cottonwood tree (Tiegs and Pohl 2005). Deep, cold water is released from dams, and this, together with all the other changes, is endangering the humpback chub in the Colorado (Petersen and Paukert 2005). Smoothing out the flow alone has altered the habitat of clams in the Colorado Delta (Cintra-Buenrostro and others 2005). Silt-free rivers create further troubles when they reach the ocean. Silt from the Nile once protected the Mediterranean delta coast from erosion, but after inauguration of the Aswan High Dam, no longer does so (Fanos 1995). Fewer nutrients now enter the Mediterranean, leaving it even more of a marine "desert" (Azov 1991).

Wetlands

Wetlands have less economic value than perennial rivers, but because their economic development value vastly increases when they are drained, they are under even more pressure. Because wetlands are the resting or roosting places for huge numbers of migratory birds, the most important have been given supposedly strong protection in international agreements, notably the Ramsar Convention of 1971, and the Convention on Biological Diversity (CBD) of 1992; yet this has done little to stem their loss (Lemly and others 2000). Desert wetlands vary hugely in size and vulnerability, and may also change substantially with management and climatic oscillations (Figure 5.11).

Lakes

Most desert lakes are dry, seasonally or for years at a stretch. The smooth surfaces of some are

Figure 5.11: The shrinking Mesopotamian wetlands

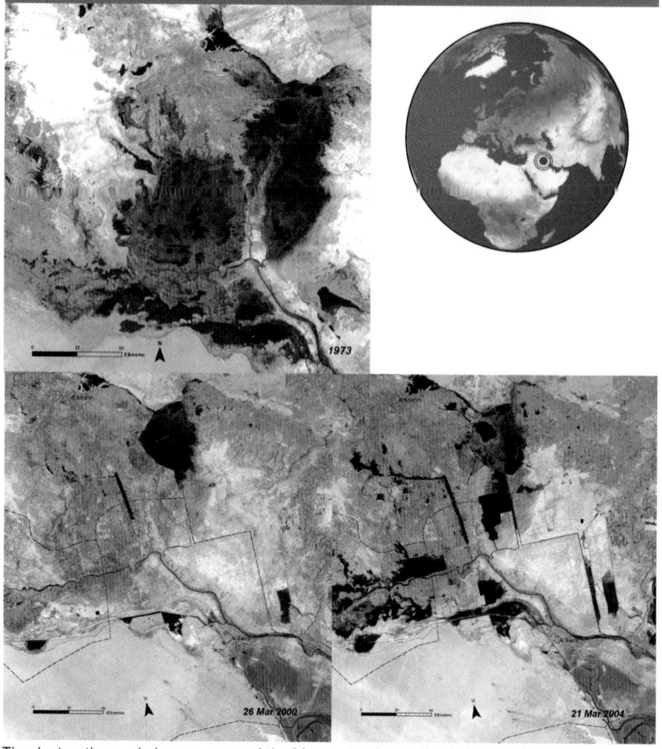

The destruction and slow recovery of the Mesopotamian wetlands in Iraq, fed by the Tigris and the Euphrates Rivers. At one time, the wetlands covered 150 000 to 20 000 km². They were still large in 1973, as the image from that year shows. However, they had lost area, even then: the first upstream dams and barrages were built in the 1950s; there are now 32, with eight more under construction and 13 more in planning. Local drainage and water diversion began in 1952. The Saddam regime sanctioned further drainage, and two-thirds of the inflow had been diverted by 1993. The 2000 image shows large brown, drained areas. The wetland once harboured two-thirds of western Asia's waterfowl; few remain. The local species of otter (*Lutra perspicialata maxwelli*), made famous in a lyrical description by Gavin Maxwell (1957) and a native turtle (*Trionix euphraticus*) are among the probable extinctions. The Marsh Arabs lost their distinctive, low-impact, much-photographed lifestyle. Now, the "Eden Project" plans to resuscitate some of the marshlands, but acknowledges that full recovery is unlikely (Kingsford and others 2005, Richardson and others 2005). The image for 2004 shows some progress: an increase in the black (flooded) area.
Source: UNEP/GRID, Sioux Falls

best-known as speed tracks, but not all are smooth. Some of their features, such as the flamingo pools in the huge Salar de Atacama in Chile, or the remarkable wind-propelled rocks at Racetrack Playa in Death Valley (Bacon and others 1996) deserve careful management, but many, like the remoter parts of the vast Salar de Atacama, and Umm-as-Samim in Oman, are protected simply by remoteness and the dangers of travel.

Lakes fed by perennial rivers, like the Aral Sea, are more biologically rich, and more vulnerable. Many have suffered severely. Owens Lake in the upper Mojave dried up in 1926 (except for a few shallow wetlands), drained by an aqueduct to Los Angeles, opened in 1913. The dry, salty lake bed now releases an estimated 900 000 to eight million tonnes of dust a year, the most prolific single source of dust in the U.S. The plume is obvious 40 km downwind (Reheis 1997). Before they were drained, lakes like this seasonally supported millions of birds. Even though saline, Lake Eyre in South Australia the world's largest ephemeral lake — occasionally supports thousands of waterfowl (Kingsford and Porter 1993). A less well-known set of lakes is probably even more biologically valuable. These are the remote, groundwater-fed lakes, some of them in extreme deserts. A few in Libya and many in China collect in hollows between huge dunes. Others are strange anomalies, like the stairway of fresh and saline lakes at Wanyanga (Ouanyanga) in extremely arid, northernmost Chad (Figure 5.12). Species endemic to isolated water bodies like these, as the desert pupfish of the Sonoran Desert, are vulnerable to extinction and need special attention (Fagan and others 2002).

Rehabilitating wet desert habitats

Given the massive size of the engineering structures that have ruined the ecology of rivers, wetlands and lakes, and the millions of people who now depend on them for water, their complete ecological restoration is unthinkable, at least in the short-term. In this time frame, better management of flow regimes, rerouting of saline return flows to special canals (as has been proposed in several irrigation schemes), and the preservation of a few

Figure 5.12: Desert lakes, Wanyanga, Chad

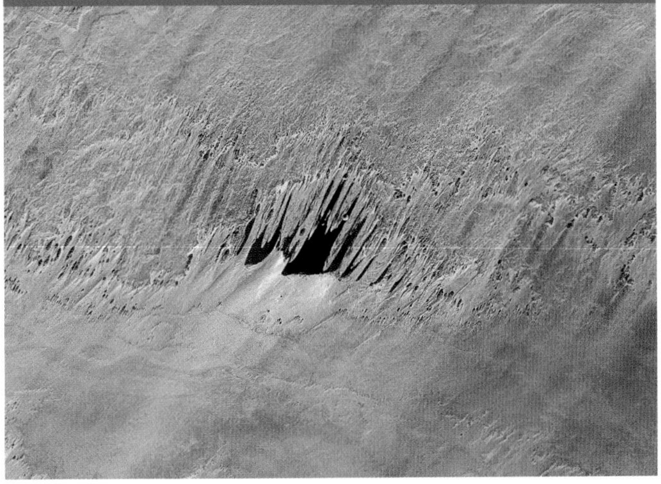

The lakes at Wanyanga (Ouanyanga) in northernmost Chad surrounded by hyper-arid desert. Dunes have invaded the depression and separated the lakes. Some of the lakes are fresh, some saline. They are isolated from other open water bodies by hundreds of kilometres of desert.

Source: Google Earth image browser

remnant wetland or lacustrine ecosystems (as is also happening in many places), would alleviate some of the more urgent conservation problems. Desalinization, as may again happen on the Colorado, is too much to hope for on a major scale. In the longer-term, the financial, human, and environmental costs of maintaining huge water-delivery systems may foreclose them. Rivers, wetlands and lakes might again return to their prelapsarian glory, but at huge human cost.

Two further wet ecosystems, both of high biological value, need rather different forms of conservation. The first is in damp desert "hollows", places to which water gravitates and feeds a shallow water-table or at best a few springs. The second are ecosystems that have been isolated by post-glacial climatic change: the "sky-islands" (see Chapter 1). In some of both, isolation has allowed the evolution of unique species, or sub-species. Unfortunately for conservationists, people (indigenous or exploitative, malevolent or innocent) also gravitate to these places, which then become the sites of intense conflict. Their conservation needs strict restrictions on interference.

DRY DESERT HABITATS

Desert ecosystems in the huge spaces between the wet places vary greatly in biological value. Some of the hyper-arid deserts support little life of any kind, except after rare rainstorms when they are briefly visited by species from surrounding, better-watered deserts. One can travel for hours in parts of southern Libya, Algeria or Peru, eastern Saudi Arabia and central Oman, northern Sudan, Chad, Niger, or northern Chile without seeing a blade of grass or a bush. A daily lizard or gazelle, or at some seasons a migrating bird, comes as a surprise. Some authorities have asked, are they ecosystems at all? Are the connections between species durable enough to class them as fully-interactive systems? (Moore 1978). It is more debatable whether these wildernesses deserve the scarce resources available to conservationists. Perhaps conservationists should concentrate their efforts on the desert's edge, that is, on arid ecosystems interposed between the hyper-arid wildernesses and the savannahs and grasslands. It is in these areas where the harvesting of wild products often offers a valuable economic alternative (Box 5.6), and more generally, where many species have great biological value and are under maximum threat.

Hunting

Indiscriminate hunting is one of the biggest threats to desert biological sustainability, so damaging that if allowed, it would soon eliminate its own *raison d'être*. Hunting may have eliminated the post-glacial mega-fauna of North America, and much later, North Africa and southwest Asia lost most of their large mammals to hunters in Roman times. It is said that on the day on which the Emperor Titus inaugurated the Roman Coliseum, 5 000 wild animals were slaughtered. The more distant African and Asian deserts lost their large mammals after guns became available in colonial times, when they were also infected with the European hunting ethos (Anderson and Grove 1987). In desert Central Asia, the native Przewalski's horse (*Equus caballus przewalskii*), was hunted to extinction in the wild by about 1870. Hunting and competition with domestic camels has reduced the population of the wild Bactrian camel (*Camelus bactrianus ferus*). Przewalski's gazelle (*Procapra przewalskii*) is now isolated to only four local populations around the Qinghai Lake in desert China (Li and Jiang 2002). Hunting continues. Income from trophy hunting in Namibia was estimated in 2003 to be 14 per cent of all tourist earnings (Humavindu and Barnes 2003). Large convoys of air conditioned caravans follow hunters across the deserts of Arabia, Sudan and Kazakhstan. The remaining populations of large mammals, such as various species of gazelle, oryx (*Oryx beisa*), addax (*Addax nasomaculatus*), Arabian tahr (*Hemitragus jayakari*) and the Barbary sheep (*Ammotragus lervia*) are on the brink of extinction. Game bird populations are declining fast, particularly that of the Houbara (*Chlamydotis undulata macqueenii*), which is the choice for hunters with falcons in Arabia and Kazakhstan (Tourenq and others 2005). Most of these endangered species need urgent protection: captive breeding (of oryx as in Oman); reintroduction; restrictive legislation and its enforcement; and so on. Surprisingly (to some) help is coming from the hunters, at least as concerns the Houbara; they have supported captive breeding, radio tracking and other conservation measures in Saudi Arabia and the United Arab Emirates (Bailey and others 1998).

Grazing

Another threat to dry ecosystems may come from pastoralists, but here the argument is not so simple. Pastoralists do indeed use the high nutritive value of arid grasslands in wet seasons. The Fulani in Niger take their cattle to the edge of the desert in the wet season, specifically for this (Penning de Vries and Djeteyé 1982), as do the Kabbabish camel nomads in Sudan (Wilson 1978).

The argument about the damage that grazing might be doing to these ecosystems is unfinished. Some authorities believe that overstocking has removed valuable species, and reduced grazing value. Their evidence includes the replacement of palatable by unpalatable species, particularly the woody shrubs that have invaded many arid areas, as in Arizona (Guo 2004) or in southern Africa (Wiegand and others 2005). Further evidence lies in research into recovery times after intense grazing. In one case, full grass cover only replaced woody scrubland after 20 years of protection (Valone and others 2002). Estimates of recovery times for the Mojave ecosystems range from 50 to 300 years, but in some cases could reach 3 000 years (Lovich and Bainbridge 1999).

The counter-argument depends on the model of these ecosystems as "pulse-response" (see Chapter 1). In seasonal or multi-annual dry periods, the ecosystem is said to be incapable of supporting enough stock to do it any damage in the brief, wet pulses. Only if stock is fed supplements brought in from wetter areas, or is moved back and forth from wet to dry areas (which, it is true, are common practices), is "overgrazing" even possible. Nomadism, this school also argues, is an ideal system for using the patchy effect of rainfall on grazing. Further, especially in the Old World, the character of many arid ecosystems depends on grazing, and because some of the native grazing animals have become domesticated, grazing by them could be vital in maintaining the original ecosystem dynamics. Conservation may then depend more on the choice of grazing intensity and grazing species, than their total exclusion. Finally, some authorities believe that overgrazing is much less of threat in arid climates where forage dynamics are primarily driven by climatic cycles, than in semi-arid or sub-humid climates where grazing is more likely to be the chief control on the availability of forage (Briske and others 2003).

Conservation priorities

Another issue in the conservation of these systems is the choice of priority. If a rare species has priority, like the long-lived desert tortoise in the arid Mojave (*Gopherus agassizii*, a federally-listed, threatened species), complete protection from hunters and graziers may be imperative. If the objective is the conservation of an ecosystem as a whole, there is more room for manoeuvre, but there are problems there too. If there are functioning patches of different size, all the way from micro-ecosystems sheltered by bushes or trees, to patches of many hundreds of square kilometres, what is the optimum size of protected areas to ensure conservation? The fail-safe rule is: as large as possible.

Community-based conservation

A more practical question is also exercising conservationists: should conservation be "expert-led" or "community-based"? (Berkes 2004). Community-based conservation does not always succeed, mainly because development and conservation goals usually conflict, and because the phrase has become too loosely used. But with care, the idea of community conservation should resonate in deserts where people and their grazing animals have been part of the ecosystem for thousands of years, and where there is huge wealth of indigenous knowledge, as Triulzi (2001) describes in Syria.

Beauty

The beauty of deserts, so essential to tourists and residents, is a further priority for conservation. There are at least three conflicting desert aesthetics: desert as wilderness, as challenge, or as scenery. The wilderness-seekers — stereotypically, backpackers, and the well-educated — use deserts to find their "intellectual humility" (Leopold 1949). To them, development and the numbers of visitors need

stringent control. The challenge-seekers need technology to achieve their aims (they might be SUV drivers). They want to drive over exciting terrain and may compromise with, even welcome, development. Given the openness of many deserts and their accessibility to motor transport, their activity can be intrusive (Sax 1980). The tourist aesthetic is subdivided many more times: the Wissa Wassef romanticism of the image on the title page of this chapter; the desert fathers (Box 5.2) and/or romantic decay; Marlboro County; walkabout; T.E. Lawrence; Rudolf Valentino; and so on. And tastes change.

These various demands need to be balanced with the needs of people who live and work in the desert, let alone with the demands of people beyond the desert for water or energy. The task is not impossible. Most visitors to the Grand Canyon or Uluru are impressed by the sensitivity with which they are managed under the onslaught of tens of thousands of people in a season, even if a minority complains one way or another. An advantage of some deserts over many other global environments is that they are remarkably resilient. Sand dunes loose the tracks of people or vehicles after a gentle breeze. Bare rock may not register a moderate number of footfalls. But the desert is not everywhere so unforgiving; there are many places where the demands clash, or where trampling, littering, and disturbance do lasting damage. Vehicle tracks in many deserts are visible on satellite images; they may persist for very many years (Belnap and Warren 2002).

All forms of conservation (particularly of biological value and beauty) are best met with some degree of statutory protection, backed by adequate funds. The trajectory of the total desert area under protection is upward, which is encouraging, but it would need to be accompanied by less easily monitored trajectories of community involvement, visitor satisfaction, biological change and funding, to confirm it as a wholly good story.

REFERENCES

Alam, U.Z. (2002). Questioning the water wars rationale: a case study of the Indus Waters Treaty. *Geographical Journal* 168(4): 341–353

Allan, J.A. (2005). Groundwater use in the Middle East and the rural transformation and environmental consequences. In: *Increasing Water's Contribution to Development in the Middle East and North Africa, MENA Regional Development Report on Water*, Report for the World Bank, Washington DC, 42 pp.

Amjad, R. (1986). Impact of workers' remittances from the Middle East on Pakistan's economy: some selected issues. *Pakistan Development Review* 25(4): 757–785

Anderson, D.M. and Grove, R. (1987). Introduction: the scramble for Eden: past present and future in African conservation. In *Conservation in Africa: peoples, policies and practice* (ed.) D.M. Anderson and R.H. Grove, pp. 1–20, Cambridge University Press, Cambridge

Azov, Y. (1991). Eastern Mediterranean — a marine desert. *Marine Pollution Bulletin* 23: 225–232.

Bacon, D., Cahill,T. and Tombrello, T.A. (1996). Sailing stones on Racetrack Playa. *Journal of Geology* 104(1): 121–125.

Bailey, T.A., Samour, J.H. and Bailey, T.C. 1998. Hunted by falcons, protected by falconry: Can the houbara bustard (*Chlamydotis undulata macqueenii*) fly into the 21st century? *Journal of Avian Medicine and Surgery* 12(3): 190–201.

Belnap, J., and Warren S.D. (2002). Patton's tracks in the Mojave Desert, USA: An ecological legacy. *Arid Land Research and Managament* 16(3): 245–258.

Berkes, F. (2004). Rethinking community-based conservation. *Conservation Biology* 85(1): 1–20

Briske, D.D., Fuhlendorf, S.D., and Smeins, F.E. (2003). Vegetation dynamics on rangelands: a critique of the current paradigms. *Journal of Applied Ecology* 40: 601–614

Bulloch, J. (1993). *Water wars: the coming conflicts in the Middle East*, Gollancz, London, 224 pp.

Catella, M., Paris, E. and Solari, L. (2005). Case study: efficiency of slit-check dams in the mountain region of Versilia Basin. *Journal of Hydrologic Engineering* 131(3): 147–154

Cintra-Buenrostro, C.E., Flessa, K.W. and Ávila-Serrano, G. (2005). Who cares about a vanishing clam? Trophic importance of *Mulinia coloradoensis* inferred from predatory damage. *Palaios* 20(3): 296–302

Cova, B. and Kerviskadic, S. (2005). *Development of cross-border interconnections as a way to promote regional energy markets*. Input 2005 Investment Conference, Accra, December

D'Amico, M.L., Navari-Izzo, F. and, Izzo, R. 2004. Alternative irrigation waters: Uptake of mineral nutrients by wheat plants responding to sea water application. *Journal of Plant Nutrition* 27(6): 1043–1059

Diamond, J. (1997). *Guns, germs and steel: a short history of everybody for the last 13,000 years*. Jonathan Cape, London 480 pp.

Diamond, J. (2005). *Collapse: how societies choose to fail or succeed*. Allan Lane, London, 575 pp.

Drake N. and Bristow, C.S. (2006). in press. Shorelines in the Sahara: geomorphological evidence for an enhanced monsoon from palaeolake Megachad, *The Holocene*

Dutkiewicz, S., Follows, M.J., Heimbach, P. and Marshall J. (2006). Controls on ocean productivity and air-sea carbon flux: An adjoint model sensitivity study. *Geophysical Research Letters* 33(2): Art. No. L02603

Evenari, M., Shanan, L. and Tadmor, N. (1971). *The Negev: The challenge of a desert*. Harvard University Press, Cambridge MA, 345 pp.

Fagan, W.F., Unmack, P.J., Burgess, C. and Minckley, W.L. (2002). Rarity, fragmentation, and extinction risk in desert fishes. *Ecology* 83(12): 3250–3256

Fanos, A.M. (1995). The impact of human activities on the erosion and accretion of the Nile delta coast. *Journal of Coastal Research* 11(3): 821–833

FAO. (2000). *FAO technical guidelines for responsible fisheries. Aquaculture Development,* 5. Food and Agriculture Organisation of the United Nations, Rome. http://www.fao.org/documents/show_cdr.asp?url_file=/DOCREP/003/W4493E/w4493e06.htm

Felger, R. and Nabhan, G. (1980). Agroecosystem diversity: a model from the Sonoran Desert. In: N.L. Gonzalez (ed.) *Social and Technological Management in Dry Lands*. AAAS Selected Symposium No.10, pp.129–149, Westview Press, Boulder, CO

Feng, Q., Liu, W., Si, J.H., Su, Y.H., Zhang, Y.W., Cang, Z.Q. and Xi, H.Y. (2005). Environmental effects of water resource development and use in the Tarim River basin of northwestern China. *Environmental Geology* 48(2): 202–210.

GAC. (2005). *Concentrating solar power for the Mediterranean region*. German Aerospace Centre, Report for the German BMU www.eid.dlr.de/tt/MED-CSP/

Giles, J. (2005). The dustiest place on earth. *Nature* 434(7035), 816–819

Glantz, M.H. (1977). Climate and weather modification in arid lands in and around Africa. In *Desertification: environmental degradation in and around arid lands* (ed. M.H. Glantz), pp. 307–337. Westview, Boulder

Gleick, P.H. (1998). *The World's Water 1998–1999: The Biennial Report on Freshwater Resources,* Island Press, Washington, D.C. (updated at www.worldwatch.org)

Gleick, P. (2001). Making every drop count. *Scientific American* 284: 40–46

Goodwin, H. (1996). In pursuit of ecotourism. *Biodiversity Conservation* 5(3): 277–291

Guo, Q.F. (2004). Slow recovery in desert perennial vegetation following prolonged human disturbance. *Journal of Vegetation Science* 15(6): 757–702

Hamidat, A., Benyoucef, B. and Hartani, T. (2003). Small-scale irrigation with photovoltaic water pumping system in Sahara regions. *Renewable Energy* 28(7): 1081–1096

Hedin, S. (1931). *Across the Gobi Desert*. Routledge, London, 402 pp.

Hirst, P. and Thompson, G. (1999). *Globalization in question: the international economy and the possibilities of governance*. 2nd edn., Polity Press, Cambridge. 318 pp.

Hosni, E. (1999). *Strategy for sustainable tourism development in the Sahara*. UNESCO United Nations Educational, Scientific and Cultural Organisation, Paris, 66 pp. http://unesdoc.unesco.org/images/0011/001196/119687eo.pdf

Humavindu, M.N. and Barnes, J.I. 2003. Trophy hunting in the Namibian economy: an assessment. *South African Journal of Wildlife Research* 33(2): 65–70

Hunt, A.G. and Wu, J.Q. (2004). Climatic influences on Holocene variations in soil erosion rates on a small hill in the Mojave Desert. *Geomorphology* 58(1–4): 263–289

Hurley S.A. and Blake A.T. 2002. The evaluation of coatings for the protection of a prestressed concrete pipeline. *Corrosion Prevention and Control* 49(1): 27–31

IPCC. (2001). *Climate Change: Impacts, adaptation and vulnerability*. Intergovernmental Panel on Climate Change, World Meteorological Organisation, Geneva, 1032 pp.

Islam, F. 2003. How these people are doing more for the Third World than Western governments. *The Observer*, London, Sunday April 20, 2003

Ives, J.D. and Messerli, B. (1989). *The Himalayan dilemma: reconciling development and conservation*. Routledge, London, 297 pp.

Jacobsen, T. and Adams, R.M. (1958). Salt and silt in ancient Mesopotamian agriculture. *Science* 128(3334): 1251–1258. [also in *Environmental geomorphology and landscape conservation*, 1, (ed. D.R. Coates), pp. 138–146, Benchmark Papers in Geology, Dowden Hutchinson and Ross, Stroudsberg, PA

Jacobson, P.J., Jacobson, K.M. and Seely, M.J. 1995. *Ephemeral rivers and their catchments: sustaining people and development in western Namibia*. Desert Research Foundation of Namibia, Windhoek, 166 pp.

Keenan, J. (2005). Waging war on terror: the implications of America's 'New Imperialism' for Saharan peoples. *Journal of North African Studies* (Special Issue — The Sahara: Past, Present and Future) 10 (3-4): 610-638

Kingsford, R.T., Lemly, A.D. and Thompson, J.R. (2005). Impacts of dams, river management and diversions on desert rivers. In: *Ecology of desert rivers* (ed. R.T. Kingsford), pp. 203–207. Cambridge University Press, Cambridge

Kingsford, R.T. and Porter, J.L. (1993). Waterbirds of Lake Eyre, Australia. *Biological Conservation* 65(2): 141–151

Kolkovsky, S., Hulata, G., Simon, Y., Segev, R. and Koren, A. (2003). Integration of agri-acuaculture systems – the Israeli experience, In *Integrated Agri-Aquaculture Systems, A Resource Handbook for Australian Industry Development* (eds. G.J. Gooley and F.M. Gavine) pp. 14-23. Rural Industries Research and Development Corporation, RIRDC Publication, Kingston, ACT, Australia, pp. 14–23

Kropotkin, P. (1914). On the desiccation of EurAsia and some general aspects of desiccation, correspondence. *Geographical Journal* 43(4): 451–458.

Kurokawa, K. (Ed.) (2003). *Energy from the desert: feasibility of very large-scale photovoltaic power generation (VLS-PV) systems*. James and James, London, 195 pp. (www.oja-services.nl/iea-pvps/products/rep8_01.htm)

Lancaster, W.O. (1981). *The Rwala Bedouin today.* Cambridge University Press, Cambridge, 179 pp.

Law, J.P. and Hornsby, A.G. (1982). The Colorado River salinity problem. *Water Supply and Management* 6: 87–103

Lemly, A.D., Kingsford, R.T. and Thompson, J.R. (2000). Irrigated agriculture and wildlife conservation: Conflict on a global scale. *Environmental Management* 25(5): 485–512

Leopold, A. (1949). *A Sand County almanac.* Sierra Club/ Ballantine, New York, 295 pp.

Li, D.Q. and Jiang, Z.G. (2002). Population viability analysis for the Przewalski's gazelle. *Russian Journal of Ecology* 33(2): 115–120

Lovich, J.E. and Bainbridge, D. (1999). Anthropogenic degradation of the southern California desert ecosystem and prospects for natural recovery and restoration. *Environmental Management* 24(3): 309–326

Lowdermilk, W.C. (1943). *Lessons from the Old World to the Americas in land use.* Smithsonian Institute, Annual Report, 415–477

Ma, J.Z., Wang, X.S. and Edmunds, W.M. (2005). The characteristics of ground-water resources and their changes under the impacts of human activity in the arid Northwest China — a case study of the Shiyang River Basin. *Journal of Arid Environments* 61(2): 277–295

Magirl, C.S., Webb, R.H. and Griffiths, P.G. (2005). Changes in the water surface profile of the Colorado River in Grand Canyon, Arizona, between 1923 and (2000). *Water Resources Research* 41(5), Art. No. W05021

Mattingly, D.J., Dore, J. and Wilson, A.I. Eds. (2002). *The archaeology of Fezzan.* Monograph 5, 2 volumes, Society for Libyan Studies, London, 461 pp.

Maxwell, G. *(1957). A reed shaken by the wind.* Longmans, London, 223 pp.

Milly, P.D.C., Dunne, K.A. and Vecchia, A.V. (2005). Global pattern of trends in streamflow and water availability in a changing climate. *Nature* 438(7066): 347–350

Moore, P.D. (1978). Are deserts ecosystems? *Nature* 275(5682): 691–692

Olivier, J. (2002). Fog-water harvesting along the West Coast of South Africa: A feasibility study *Water SA* 28(4): 349–360

Patz, J.A., Campbell-Lendrum, D., Holloway, T. and Foley, J.A. (2005). Impact of regional climate change on human health, *Nature* 438(7066): 310–317

Penning de Vries, F.W.T. and Djeteyé, M.A., ed. (1982). *La productivité des pâturages sahéliens: une étude de sols, des végétations et de l'exploitation de cette resource naturelle.* Centre for Agricultural Publishing and Documentation, Wageningen, 525 pp.

Petersen, J.H. and Paukert, C.P. (2005). Development of a bioenergetics model for humpback chub and evaluation of water temperature changes in the Grand Canyon, Colorado River. *Transaction of the American Fisheries Society* 134(4): 960–974

Plunket, P. and Urunuela, G. (2005). Recent research in Puebla prehistory. *Journal of Archaeological Research* 13(2): 89–127

Pyne, S. (1998). *How the canyon became grand.* Viking, New York, 199 pp.

Ray, I. and Williams, J. (1999). Evaluation of price policy in the presence of water theft. *American Journal of Agricultural Economics* 81(4): 928–941

Reheis, M.C. (1997). Dust deposition downwind of Owens (dry) Lake, 1991–1994: Preliminary findings. *Journal of Geophysical Research-Atmospheres* 102(D22): 25999–26008

Reisner, M.P. (1986). *Cadillac Desert: the American West and its disappearing water.* Viking, New York, 582 pp.

Richardson, C.J., Reiss, P., Hussain, N.A., Alwash, A.J., and Pool, D.J. (2005). The Restoration Potential of the Mesopotamian Marshes of Iraq. *Science* 307(5713): 1307–1311

Ritchie, J.C., Nearing, M.A., Nichols, M.H. and Ritchie, C.A. (2005). Patterns of soil erosion and redeposition on Lucky Hills Watershed, Walnut Gulch experimental watershed, Arizona. *Catena* 61(2–3): 122–130.

Rognon, P. 1991. Un projet japonais de lutte contre la sécheresse au Sahel. *Sécheresse,* 2(2): 135–138

Sax, J.L. (1980). *Mountains without handrails: reflections on the national parks.* University of Michigan Press, Ann Arbor, Michigan 152 pp.

Shafroth, P.B., Cleverly, J.R., Dudley, T.L., Taylor, J.P., Van Riper, C., Weeks, E.P. and Stuart, J.N. (2005). Control of Tamarix in the Western

United States: Implications for water salvage, wildlife use, and riparian restoration. *Environmental Management* 35(3): 231–246

Shell International (2001). *Energy needs, choices and possibilities – scenarios to 2050.* Series "Exploring the future", Global Business Environment, Shell International, London. 32 pp.

Shelley, P.B. (1818). Ozymandias. *The Examiner,* London, 1 February 1818; p.13

Silverman, B.A. (2003). A critical assessment of hygroscopic seeding of convective clouds for rainfall enhancement. *Bullletin of the American Meteorological Society* 84(9): 1219–1230

Smalley, R. E. (2005). Future global energy and prosperity: the terawatt challenge. *MRS Bulletin* 30: 412–417

Staubwasser, M., Sirocko, F., Grootes, P.M. and Segl, M. (2003). Climate change at the 4.2 ka BP termination of the Indus valley civilization and Holocene south Asian monsoon variability. *Geophysical Research Letters* 30(8): 1425

Stebbing, E.P. 1938. Africa and its intermittent rainfall: the role of the savannah forest, *Royal African Society Journal, Supp.* 37, 32 pp.

Tate, E.L. and Farquharson, F.A.K. 2000.Simulating reservoir management under the threat of sedimentation: The case of Tarbela dam on the River Indus. *Water Resources Management* 14(3): 191–208

Tiegs, S.D. and Pohl, M. (2005). Planform channel dynamics of the lower Colorado River: 1976–2000. *Geomorphology* 69(1–4): 14–27

Tiffen, M., Mortimore, M. and Gichuki, F. (1993). *More people, less erosion: environmental recovery in Kenya.* John Wiley and Sons, Chichester, 328 pp.

Tourenq, C., Combreau, O., Lawrence, M., Pole, S.B., Spalton, A., Gao, X.J., Al Baidani, M. and Launay, F. 2005. Alarming houbara bustard population trends in Asia. *Biological Conservation* 121(1): 1–8.

Triulzi-L. (2001). Empty and populated landscapes: the Bedouin of the Syrian Arab Republic between "development" and "state". *Land Reform, Land Settlement and Cooperatives* (2): 30–46

Turral, H.N., Etchells, T., Malano, H.M.M., Wijedasa, H.A., Taylor, P., McMahon, T.A.M. and Austin, N. (2005). Water trading at the margin: The evolution of water markets in the Murray-Darling Basin. *Water Resources Research* 41(7), Art. No. W07011

UNDESA. (2005). *World economic and social survey 2005.* United Nations Department of Economic and Social Affairs, New York 250 pp. http://www.un.org/esa/policy/wess/wess2005files/wess2005web.pdf

UNEP. 2002. *Global Environment Outlook 3 (GEO-3)* United Nations Environment Programme, Nairobi, 416 pp.

Uno. I., Harada, K., Satake, S., Hara, Y. and Wang, Z.F. (2005). Meteorological characteristics and dust distribution of the Tarim Basin simulated by the nesting BAMS/CFORS dust model. *Journal of the Meteorological Society of Japan* 83A: 219-239

Valone, T.J., Meyer, M., Brown, J.H. and Chew, R.M. (2002). Timescale of perennial grass recovery in desertified arid grasslands following livestock removal. *Conservation Biology* 16(4): 995–1002

Vörösmarty, C.J., Meybeck, M., Fekete, B., Sharma, K., Green, P. and Syvitski, J.P.M. (2003). Anthropogenic sediment retention: major global impact from registered river impoundments. *Global Planetary Change* 39(1–2): 169–190

Warren, A., Batterbury, S.P.J. and Osbahr H. (2001). Sustainability and Sahelian soils: evidence from Niger. *Geographical Journal* 167(4): 324–341

Wiegand, K., Ward, D. and Saltz, D. (2005). Multi-scale patterns and bush encroachment in an and savanna with a shallow soil layer. *Journal of Vegetation Science* 16(3): 311–320

Wilson, R.T. (1978). The "gizzu" winter grazing in the south Libyan Desert. *Journal of Arid Environments* 1(4): 327–344

Wittfogel, K.A. (1957). Oriental despotism: a comparative study of total power. Yale University Press / Oxford University Press, Oxford, 556 pp.

Zhang, X.-Y., Gong, S.-L., Zhao, T.-L., Arimoto, R., Wang, Y.-Q. and Zhou, Z.-J. (2003). Sources of Asian dust and role of climate change versus desertification in Asian dust emission. *Geophysics Research Letters* 30(24): Art. No. 2272

Zon, R. (1935). Shelterbelts: futile dream or workable plan ? *Science* 81(2104): 391–394

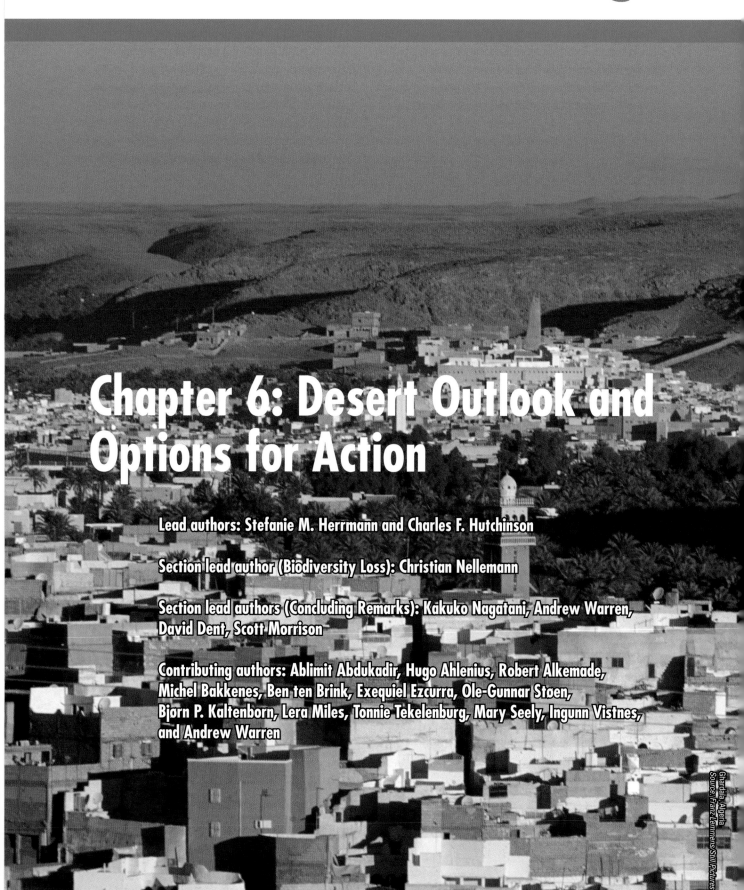

6

Chapter 6: Desert Outlook and Options for Action

Lead authors: Stefanie M. Herrmann and Charles F. Hutchinson

Section lead author (Biodiversity Loss): Christian Nellemann

Section lead authors (Concluding Remarks): Kakuko Nagatani, Andrew Warren, David Dent, Scott Morrison

Contributing authors: Ablimit Abdukadir, Hugo Ahlenius, Robert Alkemade, Michel Bakkenes, Ben ten Brink, Exequiel Ezcurra, Ole-Gunnar Støen, Bjørn P. Kaltenborn, Lera Miles, Tonnie Tekelenburg, Mary Seely, Ingunn Vistnes, and Andrew Warren

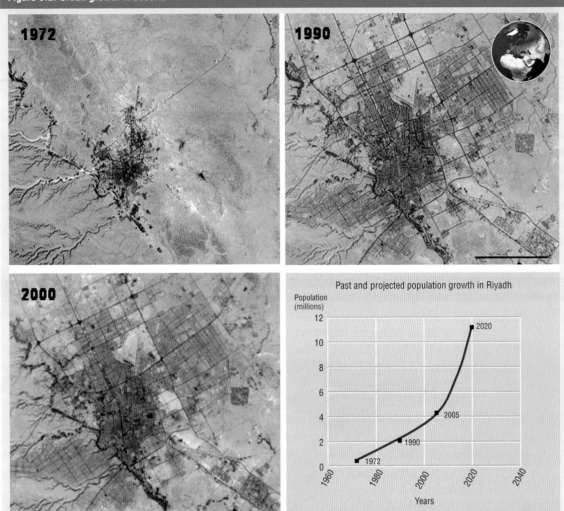

Figure 6.2: Urban growth in deserts

1972

1990

2000

Past and projected population growth in Riyadh

Population (millions)

2020

2005

1990

1972

Years

Landsat (MSS, TM and ETM+) images portraying urban growth in Riyadh, Saudi Arabia. The population of this desert city increased more than seven-fold between 1972 and 2000. The rapid growth is expected to continue into the future and the city is projected to have more than ten million inhabitants by the year 2020. Riyadh has been a major beneficiary of development interventions financed by resources resulting from the oil boom, which made it a major target for migration from rural areas and neighbouring countries.
Sources: Garba 2004, ArabNews 2005

Since the beginning of the 20th century, population in deserts of the developing world has multiplied by a factor of eight (Le Houérou 2002). Most population growth in deserts takes place in cities (Figure 6.2), where livelihood opportunities are concentrated. The proportion of urban population is already very high in those desert countries in which a modern economy has a strong secondary and tertiary sector, as in Libya (86.9%), Saudi Arabia (88.5%), Bahrain (90.2%), the United Arab Emirates (85.5%), Israel (91.7%) and Qatar (92.3%). In other desert countries, agriculture is the dominant sector and urbanization is still comparatively modest (as in Pakistan: 34.8%; Uzbekistan: 36.4%; Somalia: 35.9%; Niger:

23.3%), but growing at a rapid pace (for example, the urban population in Niger is predicted to increase by a factor of four in the next 25 years), while rural population numbers are projected to remain more or less stable (UN 2005).

Most desert populations have an age structure typical of many developing countries, with high proportions of young people. As a result of this skew towards younger cohorts, population will continue to increase, even if fertility rates decline. In contrast, the deserts of the United States, which record the highest population growth rates in the country (Sutton and Day 2004), display an opposite age structure due to the importance of retirement migration into the Sunbelt. Age structure also has important economic implications, such as labour availability, demand for education, and the size and nature of markets.

Two major demographic uncertainties in the projections of population include international migration and family structure (for example, the mother's age at first delivery, and the number and spacing of children; Cohen 2003). Movement of people is difficult to predict, as it responds to rapidly changing economic, political and environmental factors. While much environmentally-motivated migration is towards less arid and more predictable environments, people may also move into more arid areas if they offer better security, as is the case for some of the displaced people in the current Darfur conflict. By adding population to already strained infrastructure, migration can be a source of additional pressure on deserts, and make resource management more challenging. Changes within the resident population are slower, but also difficult to predict, as they result from a complex interplay of culture, society and economics. The model of demographic transition (Notestein 1945) that described the transition of populations from high birth and death rates to low birth and death rates in Europe, has less explanatory value in the developing world (Kirk 1996), where there seems to be often a predominance of female-headed households, as a result of migration of men seeking employment in more affluent areas. The long-term consequences of these family disruptions are not yet well understood but, at least initially, the process suggests a future of male-dominated urban populations, with rural populations tending toward children, the elderly, and female-headed households.

Demand for resources

As population increases, the demand for basic resources — water, food, energy, shelter — must rise, and, considering the additional effect of economic growth, the demand for these resources in deserts is expected to rise even faster than population numbers. With increasing per capita income and urbanization, consumption patterns in developing countries tend to adjust and converge with those in the industrialized world. For example, the use of domestic water in urban households in those parts of the developing world that have access to running water is significantly higher than in rural households (Roudi-Fahimi and others 2002). Economic growth also spurs energy consumption, which is notably affected by structural changes in the economy. Typically, as economies grow, they go from a prevalence of agriculture, which has a low energy demand, through a phase of energy-intensive industries to a prevalence of lighter energy-efficient industries and services, accompanied by constant increases in energy demand in the transport sector (Alcamo and others 1996).

Economic development not only imposes additional pressures on resources; it can also bring about shifts in the kinds of resources; that are demanded. Fuelwood may be replaced by higher-grade forms of energy, such as oil or gas — or even nuclear power — with different implications for the environment. Some of the resources that are scarce in deserts, such as water, food or building material, can be imported, given a certain level of economic development. This might ease the pressure on the immediate desert environment, but increases cumulative global resource demand. With economic growth, however, more resource-efficient technologies can be afforded, such as more sophisticated irrigation systems, treatment and reuse of wastewater, use of energy sources that are alternatives to local fuelwood, and the purchase rather than the household production of milk and meat. Thus, economic development not only increases demand for resources because of changes in consumption, but also holds the potential for sustainable resource management, if coupled with a favourable and stable political environment.

Climate variability and change

Over the 20th century, global average surface temperatures increased by about 0.6°C. This was the largest temperature increase in any century in the past thousand years. The warming has been attributed to anthropogenic emissions of greenhouse gases associated with forest clearance beginning in the 18th century, and the consumption of fossil fuels which accompanied industrialization in the 19th century. This last process is likely to continue through the 21st century, so that increases in atmospheric CO_2 concentrations are projected to continue (IPCC 2000). Depending on the assumed emission scenario (Box 6.1), a globally averaged temperature increase of 1.4–5.8°C is expected over the period 1990 to 2100 (Figure 6.3). Global warming has important implications for the water cycle. Increases in temperature have already driven changes in rates of evaporation and evapotranspiration, precipitation, soil moisture, water storage in snowpacks, and flow regimes of rivers. Water plays a central role in desert life: the abundance of vegetation and biodiversity are primarily governed by the availability of soil moisture; so are human livelihood opportunities,

Box 6.1: Emission scenarios

We cannot anticipate everything, but we try to assemble as many of the pieces as possible in order to predict the future. The science — or the art — of building scenarios requires a degree of control over a wide range of factors, all intricately linked. It is like a game, where we have to guess how changing one thing will affect the whole. Some elements appear simple — it is easy to imagine that rising atmospheric temperatures will melt the sea ice and cause sea levels to rise, perhaps threatening coastal populations — but at what speed and what intensity, and will this start a chain-reaction of new calamities?

To address these challenges, the Intergovernmental Panel on Climate Change (IPCC) developed four different narrative storylines for four different possible demographic, social, economic and environmental trajectories (IPCC 2000). Forty specific scenarios, each corresponding to a particular quantitative interpretation of one of the four storylines, have been developed by six modeling teams. The main characteristics of the four storylines are:

A1. The Special Report on Emissions **A1** scenario (SRES-A1) describes a future world of very rapid economic growth, a global population that peaks in mid-century and declines thereafter, and the rapid introduction of new and more efficient technologies. Specific regional patterns tend to disappear as a result of increased cultural and social interaction. The gap between regions, in terms of per capita income, reduces substantially. This scenario develops into three groups that describe alternative pathways in the development of energy supply: fossil-intensive (**A1FI**), non-fossil energy sources (**A1T**), or a balance across all sources (**A1B**).

A2. The SRES-**A2** scenario describes a heterogeneous world, based on self-reliance and preservation of local identities. Fertility patterns across regions converge very slowly, which results in a continuously increasing population, albeit at ever-decreasing rates. Economic development is regionally oriented, and per capita economic growth and technological change are more fragmented and slower than in the A1 scenario.

The four narrative scenarios of IPCC. The Special Report on Emissions Scenarios (SRES) have been constructed on two fundamental axis; one, measuring the degree of importance in decision-making given to local demands as opposed to global forces, and, the other, measuring the importance conferred to market forces in decision-making, as opposed to environmental considerations.

B1. The SRES-**B1** scenario describes a convergent world with a population that peaks in mid-century and declines thereafter (as in the **A1** scenario), but with a rapid change in economic structures towards a service and information economy, with reductions in material intensity and the introduction of clean and resource-efficient technologies. The emphasis is on global solutions to economic, social, and environmental sustainability, including improved equity, but without additional climate initiatives.

B2. The SRES-**B2** scenario describes a world in which the emphasis is on local solutions to economic, social, and environmental sustainability rather than the global approach in **B1**. It is a world with a continuously increasing global population, but at a slower rate than the **A2** scenario, intermediate levels of economic development, and slow but diverse technological change. Society is oriented towards environmental protection and social equity, and focuses on the local and regional levels.

The IPCC-SRES simulations on greenhouse gas emissions based on these four storylines result in different emission levels, which cause different radiative forcings (balances between solar radiation coming into the atmosphere and radiation going out), and as a result, different changes in global temperatures.

Box author: Christian Nellemann

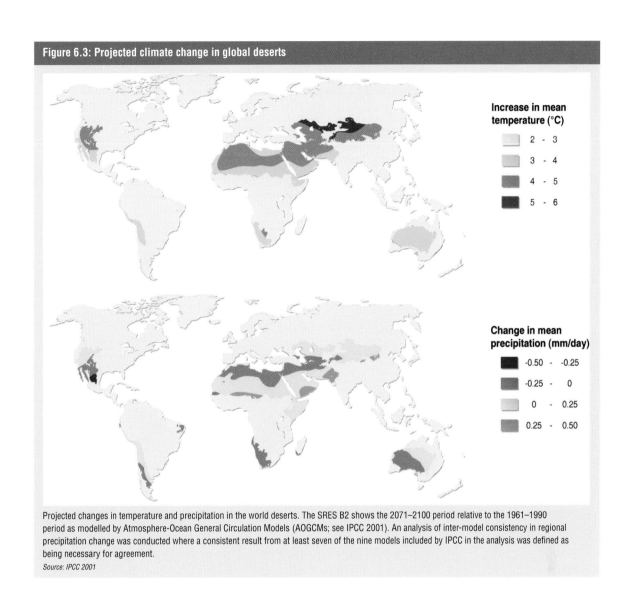

Increase in mean temperature (°C)

	2 - 3
	3 - 4
	4 - 5
	5 - 6

Change in mean precipitation (mm/day)

	-0.50 - -0.25
	-0.25 - 0
	0 - 0.25
	0.25 - 0.50

Projected changes in temperature and precipitation in the world deserts. The SRES B2 shows the 2071–2100 period relative to the 1961–1990 period as modelled by Atmosphere-Ocean General Circulation Models (AOGCMs; see IPCC 2001). An analysis of inter-model consistency in regional precipitation change was conducted where a consistent result from at least seven of the nine models included by IPCC in the analysis was defined as being necessary for agreement.

Source: IPCC 2001

directly and indirectly. Hence, it has been argued that desert environments will be very responsive to the impacts of global warming (Lioubimtseva and Adams 2004).

While all the climate models predict increases in global mean precipitation, some regions will become wetter and others drier, and there are large differences in these projections among different climate models (van Boxel 2004). For example, a study by Held and others (2005) predicted a drying trend in the Sahel over the next 50 years as a result of global warming and increases of aerosols in the atmosphere, whereas Haarsma and others (2005) expected that the warming of the Sahara might bring increased rainfall in the adjacent Sahel. Both models are global simulations, and the large differences in their

outputs result from uncertainties in the boundary conditions they adopt (such as emission scenarios) and the processes they choose to model (such as the roles of clouds, oceans, greenhouse gases in determining the disposition of solar energy). Yet, irrespective of rainfall, increases in evaporation and evapotranspiration resulting from higher temperatures will increase the potential for more severe, longer-lasting droughts in deserts.

As a general trend, there have already been reports of increases in the variability of rainfall and in the frequency of extreme events (Salinger 2005). Interannual rainfall variability caused by the El Niño Southern Oscillation (ENSO; see Chapters 1 and 3) and North Atlantic Oscillation (NAO) cycles is likely to increase further, which will reinforce the pulse and reserve dynamics governing desert

Increasing aridity, as a result of complex repercussions of 21st century global warming, could turn some semi-arid rangelands into deserts and cause the re-mobilization of relict dunes that are currently stabilized by vegetation.

Simulation experiments by Thomas and others (2005), based on a variety of global climate change models, suggest that some of the stabilized dune systems in the southern African Kalahari basin could be reactivated by way of decreases in soil moisture and increases in the number of extreme climatic events. Both factors boost the 'dune mobility index', a measure of potential sediment mobilization combining surface erodibility (vegetation cover and soil moisture), and atmospheric erosivity (wind energy). Independent of the specific climate change model that these authors used, their simulations predicted significant enhancement of dune mobility in the southern Kalahari dune fields by 2039, and in the eastern and northern dune fields by 2069.

Other relict dune fields, which could potentially experience climate-induced re-mobilization in the coming decades, include Australian and North American dunefields (Knight and others 2004).

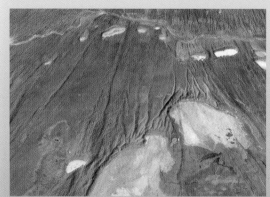 

Dunes of the southern Kalahari near the Botswana/South African border. Most of this dune landscape is, at present, stable with only limited sand movement on the crests of the ridges (red areas) and thin vegetation cover in the interdune areas (greenish - brown areas). As climate change proceeds, this whole landscape may become mobile in a few years.

Source: Landsat ETM+ true colour image from September 2001 draped over SRTM (Shuttle Radar Topography Mission) elevation data, GLCF, University of Maryland

Box author: Stefanie M. Herrmann

ecosystems, triggering potentially fewer but more intense biologically significant rainfall pulses. There is evidence that higher drought incidence is likely to reinforce, or at least expose, desertification/degradation processes (Le Houérou 1996), such as permanent losses of bioproductivity and biodiversity, erosion and deflation — and could lead to the spread of some processes that have been assumed to be under control, such as the re-mobilization of vegetated sand dunes (Box 6.2).

Some of the water in some deserts comes from rivers that originate outside the desert boundary, often in the snow and ice packs of high mountains. For example, the Colorado River, which brings water to the arid American Southwest, is fed by summer snow-melt in the Rocky Mountains, and the Central Asian deserts receive water from rivers which rise in the Central Asian mountains (see Box 6.3). Ice and snow in these mountains constitute an important reservoir of water, which slowly releases water during the summer months.

Global warming has already reduced the thickness and extent of snow packs and caused seasonal shifts in stream-flow. The projected increases in temperature over the coming decades will have serious impacts on the hydrological cycle and regional water supply, by affecting accumulation and duration of snow cover, rate of melting and long-term water storage in glaciers (Barnett and others 2005). By way of global atmospheric teleconnections, changes in pressure systems in different parts of the globe can have hydrological implications for deserts. Thus, Archer and Fowler (2004) found a significant relationship between the variability of the North Atlantic Oscillation and winter precipitation in the Karakorum, which can be useful in predicting summer run-off in the Indus basin.

Globalization

Globalization, defined as the increasing worldwide integration of markets for goods, services, labour, and capital, is a major driving force of economic and environmental change, with potentially dramatic and unforeseeable impacts on development and environmental change in deserts. The past few decades have been characterized by a general shift from protectionism and state-dominated economies to freer trade and privatization; from local and national-scale economic activities to increased international flows of capital, information and goods; and from a strict dependence on local natural resources to a growing importance of technology, infrastructure and institutions for development (Di Castri 2000). Globalization has moved forward unevenly; the geopolitical opening that coincided with the end of the Cold War, the economic and financial openings defined in the General Agreement on Tariffs and Trade, and the opening of a global information society with the establishment of the Internet, have all greatly accelerated global homogenization and interconnectedness. In response, a counter-trend of increased cultural diversification, revival of local languages, and indigenous identities has sprung up in many places.

The temporal and geographic patchiness of globalization make projecting its future extremely difficult. However, the implications for the causation, or solution, of environmental problems are enormous and are frequently overlooked in purely environmental studies. On the one hand, globalization (especially the spread of free market economies) has the potential to increase economic disparities and widen social gaps within and among countries. A further marginalization of desert economies would have grave consequences for the environment, as poverty forces people to forego proactive and sustainable resource management in the pursuit of immediate survival (Panayotou 2000). But globalization can also mean new opportunities to enhance economic development while improving environmental conditions. Although environmental issues, particularly the lack of water, will continue to play a role in development if globalization proceeds, they will likely become less decisive than the economic and human factors that will be increasingly mobilized to overcome them, such as innovation, infrastructure, and marketing of assets and available resources. Diversification of the economy

that reduces reliance on subsistence agriculture on marginal lands might follow from the enlargement of markets and new marketing opportunities, easing pressures on resources and environment.

SCENARIOS OF CHANGE FOR WATER AND LAND DEGRADATION

As alternative images of the future, scenarios assist in understanding how complex systems might perform. Scenarios are not meant to be predictions or forecasts; many variables and inter-relationships within and among natural and social systems are insufficiently understood, so that precise predictions are not possible. Uncertainties arise from the quality of the data that are used, the incomplete understanding of the functioning of a system, and approximations and generalizations made in the scenario-building process (IPCC 2000). Nevertheless, scenarios are useful tools for scientific assessment and for policy-making because they provide a focus for discussion, help to organize statements about the future, and point out critical trends that could jeopardize sustainable development (Raskin 2005). Scenarios generally have a narrative component in the form of a story, and a quantitative component represented by a numeric model that may illustrate and support the story. Some systems are well understood and can be supported by appropriate quantitative data; others are better communicated by descriptive stories. In practice most scenarios combine both (Kok and van Delden 2004).

Water

Water is a critical resource for human development and environmental health, particularly in deserts. Many desert countries face serious water shortages. According to thresholds proposed by Falkenmark and Widstrand (1992) on the basis of water required to maintain an adequate quality of life, a country or area whose renewable freshwater availability drops to less than 1 700 cubic metres per person per year experiences water stress. Water scarcity, a condition in which chronic water shortages affect human and ecosystem health, and hamper economic development, occurs when the renewable freshwater availability falls below 1 000 cubic metres per person per year (Figures 6.4 and 6.5).

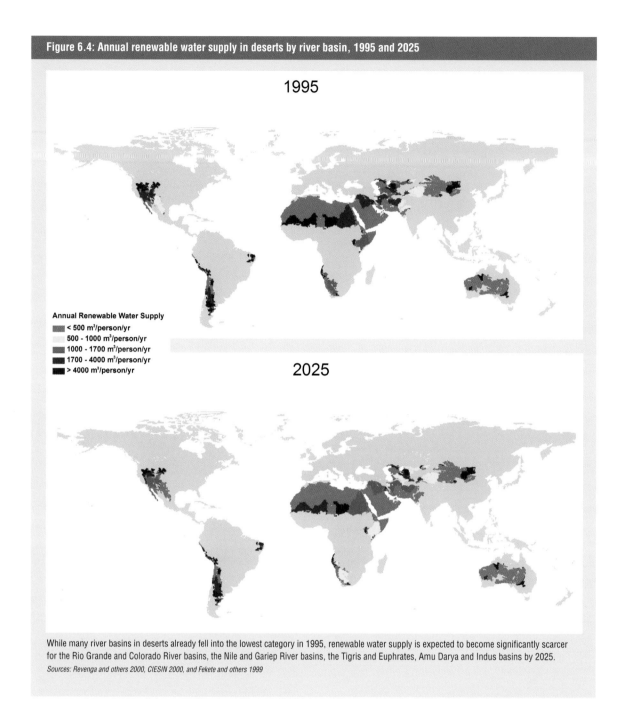

1995

Annual Renewable Water Supply
- < 500 m³/person/yr
- 500 - 1000 m³/person/yr
- 1000 - 1700 m³/person/yr
- 1700 - 4000 m³/person/yr
- > 4000 m³/person/yr

2025

While many river basins in deserts already fell into the lowest category in 1995, renewable water supply is expected to become significantly scarcer for the Rio Grande and Colorado River basins, the Nile and Gariep River basins, the Tigris and Euphrates, Amu Darya and Indus basins by 2025.

Sources: Revenga and others 2000, CIESIN 2000, and Fekete and others 1999

Future water availability is a function of future supply and demand. Water supply is controlled by climate and water demand is driven by demographics and economic factors (Arnell 1999). Climate change has already effected changes in the global water cycle, and even larger changes are projected as global warming continues (IPCC 2001). Subtle shifts in mean temperature and precipitation have brought about important changes in the occurrence of extreme climate events. Deserts and desert margins are particularly vulnerable to soil moisture deficits resulting from droughts, which have increased in severity in recent decades and are projected to become even more intense and frequent in the future. Conversely, flood events are predicted to be fewer but more intense, in which case little moisture would infiltrate into soils and run-off and eroded sediment would concentrate in depressions, reinforcing the patchiness of the desert ecosystem.

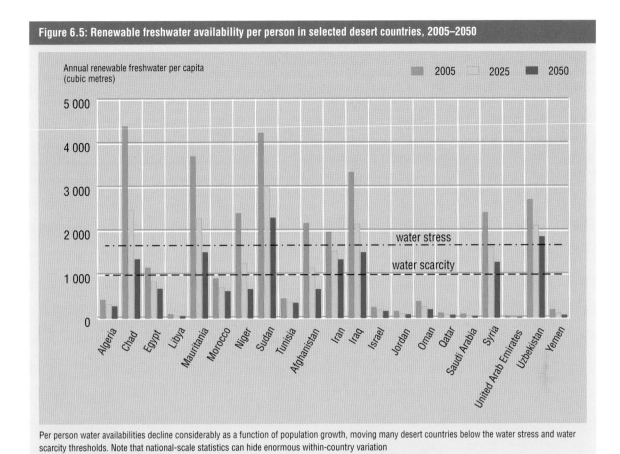

Figure 6.5: Renewable freshwater availability per person in selected desert countries, 2005–2050

Annual renewable freshwater per capita (cubic metres)

■ 2005 ▢ 2025 ■ 2050

water stress

water scarcity

Per person water availabilities decline considerably as a function of population growth, moving many desert countries below the water stress and water scarcity thresholds. Note that national-scale statistics can hide enormous within-country variation

Sources: Gleick 2006, and UN 2005

Climate change will likely affect the total amount of available water less than it will the overall water regime and the timing of water availability in deserts — particularly deserts whose water supply is currently provided by melting snow or ice. Thus, a large fraction of the water used for agricultural and domestic purposes in the arid Southwest of the United States, the deserts of Central Asia, and the Atacama and Puna Deserts on both sides of the Andes, is drawn from rivers that originate in glaciated/snow-covered mountains. As the volume of snowpack diminishes, river regimes change from glacial to glacio-pluvial and then to pluvial. As a result, total run-off is expected to increase as the glaciers begin to melt and then to decrease as the total area covered by snow and ice declines (see Box 6.3, and Yao and others 2004). Peak discharges will shift from the summer months, when the demand is highest, to the spring and winter, with potentially severe implications for agriculture. Climate and stream-flow scenarios

estimate that California's irrigated farmlands are likely to lose more than 15 per cent of their value because of losses in snowpack (Service 2004).

Water demand of the natural environment is likely to grow as potential evaporation increases as a result of warming. Increases in potential evaporation are projected to reach 7.5–10 per cent by 2020 and 13–18 per cent by 2050, depending on the global scenario used (Arnell 1999). A more important factor in the increase in water demand in deserts, which is harder to quantify, is their growing population and its aspirations for an improved standard of living. Water demand will increase rapidly in some desert areas, particularly the expanding urban areas, and only moderately in others. However, water-use per person has been rising less rapidly than previously predicted and is actually declining in a few parts of the world thanks to improvements in water-use efficiency in the agricultural, municipal, and industrial sectors.

As one of the largest glaciated areas on earth, the Tibetan Plateau together with the surrounding Himalayan mountains plays an important role in freshwater storage and provides water to the adjacent Central Asian deserts in the north (Kara-Kum, Kyzyl-Kum, Taklimakan) and the Indus Valley in the south by way of rivers fed by glacier run-off. Irrigated agriculture along the Amu Darya, Syr Darya and Tarim rivers in the Central Asian deserts as well as in the Indus basin depends heavily on water originating in High Asia (Winiger and others 2005).

Böhner and Lehmkuhl (2005) estimated the current extent of glaciers in High Asia from aerial photographs and satellite images, and modeled the potential decreases of glacial area by predicting the magnitudes of climate change under alternative socio-economic emission scenarios (IPCC 2000). Their results showed small changes in precipitation but considerable increases in annual mean temperature for High Asia (1°–1.8°C by the end of the 21st century in the best-case scenario and 3.8°–6°C in the worst-case scenario), which exceed global average rates of increase. Given the high sensitivity of the snow and ice cover to warming, the projected temperature increase has severe consequences for the glaciated area, which would decrease by 42.5 per cent in the best-case and by 81.4 per cent in the worst-case scenario. In the latter case, continuous ice cover would be limited to the Himalaya, Karakorum, Pamir and West Kunlun, whereas the Qinling Shan and Tian Shan mountain ranges would be left almost without glaciers. In addition, the slow thawing of permafrost areas and their transformation into peat bog in the more humid areas might in turn enhance global warming as a result of methane emissions.

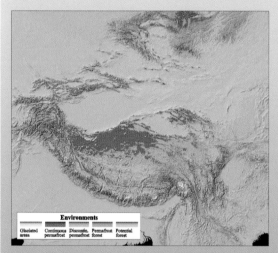 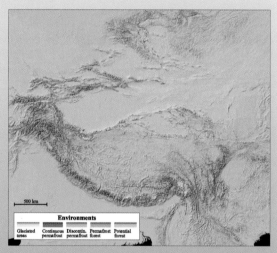

Modelled recent distribution (1961-1990) of climate-controlled environments in High Asia [left] and their potential future distribution (2070-2099) based on the A2-SRES-scenario (see Box 6.1) [right]. The extent of glaciated areas and permafrost areas is projected to diminish considerably by the end of the century under this emission scenario.

Reproduced from Böhner and Lehmkuhl 2005 by permission of Taylor & Francis AS

Box author: Stefanie M. Herrmann

Despite this positive development, there is concern about whether improvements in water use efficiency will keep pace with the projected growth in population (Gleick 2001, 2006).

Due to a shortage of surface water resources, many desert countries rely heavily on the exploitation of groundwater. For example groundwater currently provides for 95 per cent of Libya's freshwater needs and 60 per cent of Algeria's (UNEP 2002a). Most deep groundwater extracted in deserts was put in place thousands of years ago under wetter climatic conditions during the Pleistocene and is considered non-renewable on a human timescale (see Chapter 5). With the number of deep wells increasing exponentially in many areas, groundwater

has been extracted at a large scale over the past five decades. While some reserves are estimated to be vast and likely to last for a long time at current rates of exploitation, others are being depleted rapidly and are already experiencing declines in water levels and water quality (Moench 2004).

Quantification of groundwater resources is extremely complex, particularly in the absence of reliable information on groundwater extraction. The only systematic global-scale groundwater survey was compiled by the United Nations (UN 1990) and has not been updated since. The lack of precise information on groundwater availability and recharge rates poses a major challenge to sustainable management of the resource. Problems of groundwater exploitation have become more acute and more widespread under the pressures of population growth and urbanization, exacerbated by growing competition between various sectors (Vörösmarty and others 2000). Demands on water by municipal and industrial uses are expected to increase at the expense of irrigated agriculture; for example, water transfers from agricultural regions previously supported by Colorado River water have already become a common means of addressing water shortages in urban southern California (Johns 2003).

In addition to the limited quantity of water resources available in deserts, deterioration of their quality is another concern. Because of their dependence on dwindling water resources, societies in deserts are particularly vulnerable to the effects of water pollution, which threaten human and livestock health and socio-economic development. The degradation of both surface and groundwater resources by agrochemicals, mostly pesticides and fertilizers used in irrigated agriculture and the salinity of return flow (Chapter 5), is likely to increase in the future, if the expansion of irrigated lands continues without any significant improvements in drainage and treatment of agricultural wastewater. Groundwater quality often deteriorates where extraction levels are high — and these are projected to increase, particularly in fast-growing urban areas — because of the inflow of more saline deep groundwater or seawater in coastal desert areas. Future seawater intrusion into groundwater may also be caused by sea level rises resulting from global warming (IPCC 2001).

Land degradation

Land degradation is arguably one of the major global environmental challenges. Although its precise definition has stirred debate — even more so in the definition of desertification — land degradation occupies a prominent place in major environmental conventions and initiatives (among them the United Nations Convention on Environment and Development, the United Nations Convention to Combat Desertification, the World Summit for Sustainable Development, and the Millennium Ecosystem Assessment). In one of the more authoritative definitions, the UNCCD defines land degradation as "the reduction or loss of the biological and economic productivity and

complexity of terrestrial ecosystems, including soils, vegetation, other biota, and the ecological, biogeochemical and hydrological processes that operate therein … resulting from various factors including climatic variations and human activities" (UNCCD 1994).

Deserts in the strict sense are less susceptible to land degradation than other ecosystems for two reasons: (1) their biological productivity is very low; and (2) vast desert areas are almost devoid of human population, and human impact. However, desert margins, oases and irrigated lands within deserts have a higher biological potential and are subject to increasing population pressure, and thus tend to constitute potential hotspots of degradation.

As a creeping environmental problem with low-grade, incremental changes over time (Glantz 1994), land degradation is difficult to measure with any level of precision and this is one explanation of the widely diverging estimates of the extent and severity of the problem. The Global Assessment of the Status of Human-Induced Soil Degradation (GLASOD), commissioned by UNEP in 1988 as the first comprehensive soil degradation overview at the global scale, estimated the extent of highly to very highly degraded soil in deserts to be around nine per cent — considerably less than in semi-arid drylands. Other studies maintain that degradation is not as prevalent a phenomenon in drylands as suggested by many global-scale assessments, which often suffered from subjectivity, poorly representative ground data, and poor resolution. Rather, degradation seems to be concentrated in specific locations, such as around settlements and boreholes (see for example, Warren 2002).

In oases, soil salinization and the encroachment of sand dunes are major problems. Soil salinization occurs in two ways: (1) intrusion of saline seawater into deep coastal aquifers — such as the decline of oases on the coastal plain of Batinah in Oman (Stanger 1985) — a rather minor issue, though, on the global scale, but locally significant; and (2) evaporation of excess irrigation water, often associated with poor soil drainage, that leaves dissolved salts in the soil — a widespread problem in deserts globally (see Chapter 5). Sand dune encroachment into oases, a recurrent and normal phenomenon, can be exacerbated by degradation of the vegetation cover on surrounding pastures (resulting from prolonged drought or overgrazing), which exposes sandy soils to deflation. Along the Wadi Draa in southern Morocco, sand has moved into irrigation channels and palm groves (Corsale 2005).

Given the difficulties in estimating the current status and extent of land degradation in deserts, making projections into the future is uncertain. Land degradation is a complex phenomenon, which is affected by changes in a number of human and environmental factors, the projections of which are themselves error-prone: population numbers, resource demand, climate, trade and technology, and political/institutional factors being foremost among them. Furthermore, we have only incomplete knowledge about ecological thresholds to degradation and recovery potential of once degraded lands, which vary depending on their soil and geomorphic age (Brown 2000).

Few studies have offered a future outlook for land degradation. One is the "2020 Vision for Food, Agriculture and the Environment", an ongoing initiative by the International Food Policy Research Institute (IFPRI) aimed at developing a shared vision on how to meet future world food needs while reducing poverty and protecting the environment. The report expects a reduced expansion of irrigated area by the year 2020, and increased investment in drainage to deal with salinization. Nevertheless, they believe that problems of salinization will increase, as irrigation systems with inadequate drainage continue to age. Potential hot spots for this kind of soil degradation in deserts include the Nile delta, the Indus, Tigris and Euphrates alluvial lands and parts of northern Mexico (Scherr 1999). On the other hand, a considerable amount of unsustainable irrigated land is projected to go out of production and new opportunities for rehabilitation of degraded lands and sustainable pasture management systems are expected to be developed for them.

Biodiversity Loss

WHAT IS BIODIVERSITY LOSS?

Biodiversity is a broad and complex concept. It encompasses the overall variety found in the living world, from genes, to species, and ecosystems. Here, we will focus on species, considering the variety of plant and animal species in a certain area (species richness) and their population sizes (species abundance). Population size is the number of individuals per species, generally expressed as the abundance of a species or briefly "species abundance". The various types of natural regions in the world, also called "biomes", vary greatly in number and composition of species; a tropical rainforest is entirely different from deserts or tundras.

The markedly increased rate of biodiversity loss we face today is the unintentional result of increasing human activities all over the world. The process of biodiversity loss is generally characterized by a decrease in the abundance of many original species and an increase in abundance of a few other, opportunistic species as a result of human activities. Extinction is the last step of what may be a long process of degradation; numerous local extinctions may precede the potentially final global extinction. As a result of human development, many ecosystems that had differed from one another are becoming increasingly alike, more "homogenized". Decreasing populations are as much a signal of biodiversity loss as strongly expanding species, which may sometimes become invasive.

Although we know that global biodiversity is declining at an unprecedented rate (Jenkins 2003), until recently it was difficult to measure the process of biodiversity loss, as species richness appeared to be an insufficient indicator. First, it is hard to monitor the number of species in an area, but more importantly, disturbance may initially increase species richness as original species are gradually replaced by new human-dispersed invasives. For this reason, the Convention on Biological Diversity (CBD) has chosen to use, among other measures, species abundance as an indicator of biological degradation. Thus, in this report and in the GLOBIO model, biodiversity is defined as the whole set of original species and their corresponding abundance. Furthermore, because of the effort involved, thorough mapping and monitoring across large areas is not feasible in most regions. Fortunately, there are numerous thorough peer-reviewed empirical studies that quantitatively link changes in habitat, such as fragmentation, to biodiversity loss. Based on literature for specific habitat types and the extent of the pressures present, we can model the potential loss in biodiversity compared to the undisturbed state by projecting the impact of changes in different pressures over time. By comparing and analyzing historical changes in habitats, including use of satellite imagery, records in changes can be projected using different types of scenarios and assumptions.

Biodiversity loss is expressed here as the average species abundance of the original species compared to the natural or low-impacted state. Increasing exotic populations do not compensate for the loss of decreasing populations in the indicator. If the indicator is 100 per cent then the biodiversity is similar to the natural or low-affected state. If the indicator is 50 per cent then the average abundance of the original species is 50 per cent of the natural or low-affected state, and so on. To avoid masking, significant increased populations of original species are truncated at 100 per cent, although they should have a negative score.

MODELLING BIODIVERSITY CHANGE IN DESERTS

Numerous models have been developed across the last decade to assess biodiversity change (Melillo 1999, Alward and others 1999, Sala and others 2000, Clausenn and others 2003, Potting and Bakkes 2004). In 2004, several major global models merged in the creation of the GLOBIO 3.0 model, developed to meet the requirements of the CBD for a model suitable for simulating the rate and extent of biodiversity loss by estimating abundance of selected species, and to conduct projections of change in land use, infrastructure development, climate, and pollution. The individual components of the GLOBIO 3.0 model have been used to generate global scenarios (UNEP 2002b, CBD 2006), scenarios for the Arctic (Nellemann 2001, Ahlenius and others 2005), rainforest habitat for great apes in Asia and Africa (Nellemann and Newton 2003, Caldecott and Miles 2005, Butler 2003), and mountain regions (Blyth and others 2002, Nellemann 2005). The model results presented here use a definition of the desert biome

described in Chapter 1, that is, an area of particular aridity, eco-regional and land-cover attributes.

By using the SRES scenarios (see Box 6.1 for details), which have previously been used to describe the implications of different socio-economic developments for climate change, four different biodiversity scenarios have been developed for the desert biome. In this report, we present the results for biodiversity of the A2 scenario, or "regionally-defined market force" scenario. This scenario assumes that market forces will continue to be the main drivers of natural resource use, but that globalization is likely to reach a limit, giving way to renewed emphasis on local economies. We run the simulations for this scenario on the desert biome. The results of the other alternative scenarios are presented in parentheses as a range to provide an indication of the level of uncertainty. However, the A2 scenarios must not be considered a worst-case situation, but as the scenario closest to the trends of development as we have known them in the previous decades with regard to land degradation and subsequent impact on biodiversity and human livelihoods. It is also important to emphasize that because of the time lag for measures to control environmental degradation to take effect, all scenarios, even those with outstanding new efforts, will slow but not stop the rate of biodiversity loss.

Precipitation and temperature patterns likely to change

According to the SRES A2 scenario, great changes in both rainfall and temperature (and subsequent growth conditions) may take place across the world's deserts (Figure 6.3). The impacts will be highly variable from one region to the next, but they are likely to be felt the hardest in desert margins and in montane areas, as these are where the primary arid rangelands are located. Because deserts are driven by climatic pulses more than by average conditions, even moderate changes in precipitation and temperature may create severe impacts by shifting the intensity and frequency of extreme periods, and with perhaps catastrophic effects on the viability of plants, animals, and human livelihoods.

Land use intensifies in desert margins

Agricultural development, including irrigation, croplands, and grazing, is generally concentrated in oases, rangelands at desert margins, or in the lower slopes of montane desert areas. Great changes have taken place on the margins of virtually all desert areas in the world, particularly by grazing, over the last 150 years (Goldewijk 2001, Loreau and others 2001, Tilman and others 2001). According to the scenario, while expansion of croplands into deserts will be limited — except where fueled by irrigation — grazing by livestock and cutting of firewood will continue to increase inside deserts in montane areas, as well as on the desert margins (Figure 6.6).

Piecemeal development of sky-islands and desert margins

The changes and effects of land use both on the desert margins and inside the desert regions are deeply influenced by, and reflected in, piecemeal development of transportation networks, which are necessary for accessing, developing, and transporting people, goods and services, and for agricultural and livestock expansion (Leinbach 1995). Infrastructure development has been shown to disrupt the physical environment, alter the chemical environment, impact species relationships, accelerate introduction of invasive species, modify animal behavior and change land use near developed roads (Andrews 1990, Forman and Alexander 1998, Trombulak and Frissell 2000, Nellemann 2001). Desert wilderness areas (any area located more than 5 km from any infrastructure), are projected to decline from 59 per cent of the total desert area in 2005 to a low 31 per cent by 2050 (range 31-44 %), suggesting a relative loss of nearly half of the remaining intact wilderness within a few decades. This decline will primarily affect the more productive areas in desert margins and in montane areas, while the wilderness areas that remain will be primarily confined to barren areas with very low biodiversity, and where human settlements or development are not possible (SRES A2; Figure 6.7). Populations of desert bighorn sheep (*Ovis canadensis*), desert tortoises (*Gopherus agassizii*), and many species of birds have been shown to be very sensitive to fragmentation of habitat by roads (Bleich and others 1990, Edwards and others 2004, Epps and

others 2005, Gutzwiller and Barrow 2003). The same applies to many species of antelope, which are vulnerable to poaching concentrated along road corridors (Nellemann 2005), or Asian houbara bustards (*Chlamydotis macqueenii*; Bekenov and others 1998, Spalton and others 1999, Combreau and others 2001, Mesochina and others 2003).

Desert sky-islands and wetlands in alarming decline

While the wilderness areas in hot deserts decline in the model by up to about 0.8 per cent every year as a result of human development and disturbance, the change in the desert margins is much greater (Figure 6.8). Here, relatively pristine natural rangelands may decline by 1.9 per cent annually. Wetlands are at even greater risk, as they are being drained for irrigation and agricultural expansion (see Chapter 5). Of greatest risk are the few patches of forest and woodlands associated with desert montane areas and the relatively moister desert margins, or riparian habitats next to settlements (Bleich and others 1990, Bekenov and others 1998, Spalton and others 1999, Combreau and others 2001, Mesochina and others 2003, Gutzwiller and Barrow 2003, Zhao and others 2004, Epps and others 2005, Nellemann 2005). These areas are important not only for biodiversity, but are targeted because of water resources, potential pastures for livestock, and are also subject to cutting for firewood — a scarce resource in drylands. Pristine woodlands in deserts, such as montane habitats, may decline by up to 3.5 per cent annually, especially at lower elevations. This is particularly alarming, as the vegetation may be essential for reducing erosion, and logging may increase sediment loads in rivers, reduce water quality, and increase the risk and severity of flash floods (Nellemann 2005).

Currently, deciduous forests and needle-leaf forests cover only 0.13 and 0.68 per cent of the desert biome in isolated sky-island patches respectively, but represent major biodiversity hotspots at risk, as they are frequently targeted for development. Historically, montane areas have been important for cross-desert transport and settlements because they constitute important water sources (see chapter 4). Wetlands in deserts are even rarer, occupying less than 0.01 per cent of the biome.

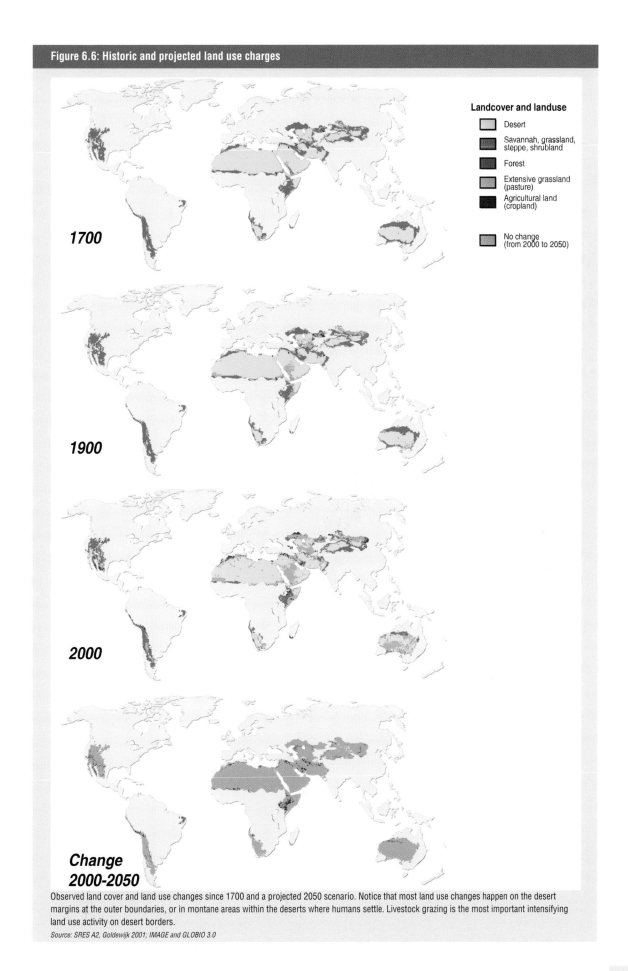

Landcover and landuse

Desert

Savannah, grassland, steppe, shrubland

Forest

Extensive grassland (pasture)

Agricultural land (cropland)

No change (from 2000 to 2050)

1700

1900

2000

Change 2000-2050

Observed land cover and land use changes since 1700 and a projected 2050 scenario. Notice that most land use changes happen on the desert margins at the outer boundaries, or in montane areas within the deserts where humans settle. Livestock grazing is the most important intensifying land use activity on desert borders.

Source: SRES A2, Goldewijk 2001; IMAGE and GLOBIO 3.0

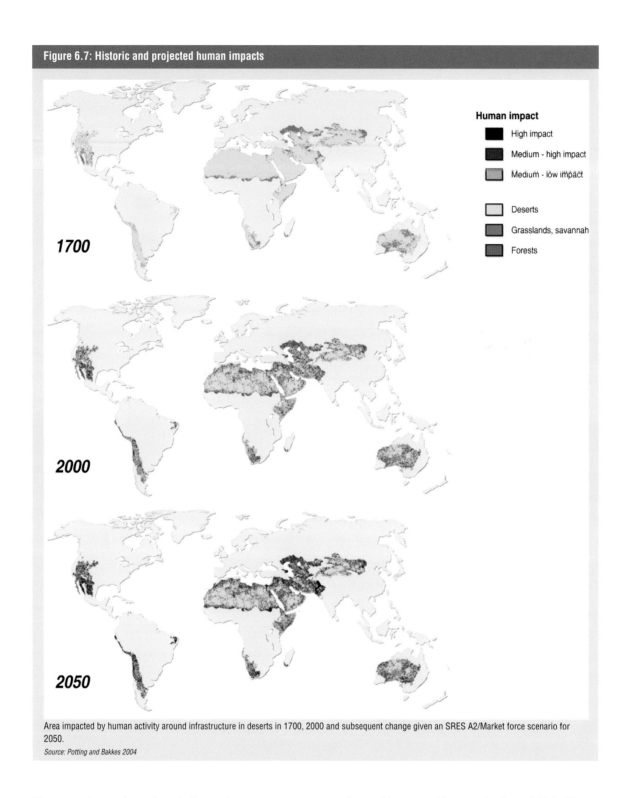

Figure 6.7: Historic and projected human impacts

Human impact
- High impact
- Medium - high impact
- Medium - low impact

- Deserts
- Grasslands, savannah
- Forests

1700

2000

2050

Area impacted by human activity around infrastructure in deserts in 1700, 2000 and subsequent change given an SRES A2/Market force scenario for 2050.

Source: Potting and Bakkes 2004

They are also projected to decline at fast rates, mainly due to drainage for irrigation and cropland development. All these habitats form hotspots of biodiversity and exhibit a very patchy distribution (Hernández and Bárcenas 1996, Riemann and Ezcurra 2005), restricted to mountainous regions (Hernández and others 2001), or to riparian zones, wadis, and in oases (Zhao and others 2004). The vulnerability of these habitats is mostly due to their isolation and fragmentation, lack of opportunity for migration of the biota when conditions change, limited extent and high endemism, and locally-restricted species that are particularly vulnerable to change (Ezcurra and others 2001).

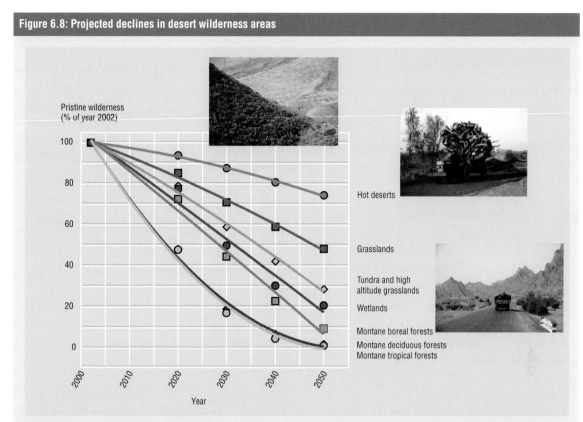

Projected declines in wilderness areas among different land cover categories within the desert biome as a result of increased infrastructure development and subsequent land use. Notice that montane areas, primarily because of their ability to supply food, are at the greatest risk of further infrastructure development and intensified land use, particularly the forests located in the foothills of montane areas within deserts, and dry rangelands at the desert margins.

Source: GLOBIO 2.0

Apart from providing important resources for livestock grazing, montane habitats are also crucial for water supply to the surrounding deserts (Table 6.1); cutting of their forests greatly diminishes the ability of the mountains to regulate water flow. Hence, casualties are not uncommon downstream from flash floods exacerbated by unsustainable land practices (Nellemann 2005). Unfortunately, only a fraction of these montane habitats is protected.

Assessing biodiversity loss in deserts

During the 6th meeting of the Conference of Parties of the CBD, the parties committed themselves to "achieve by 2010 a significant reduction of the current rate of biodiversity loss at the global, regional and national levels" (CBD 2002). Until recently, there was little quantitative data available on recent changes in species abundance, and most studies relied extensively on expert or

Table 6.1: An overview of the major rivers in Central Asia originating in mountains and flowing through or to deserts and drylands

River	Basin area (km² × 10³)	Population /km²	Total population (millions)	Water m³/person /yr	% cropland	% irrigated	% forest	% original lost	% basin protected	Dams in basin*	Hydrological significance of mountains
Tarim	1 152	7	8.1	754	2.3	0.6	<1	69	20.9	-	Very high
Syr Darya	763	27	20.6	1 171	22.2	5.4	2.4	45	1.0	7 (11)	Very high
Amu Darya	535	39	20.9	3 211	22.4	7.5	0.1	98.6	0.7	6 (8)	Very high
Indus	1 082	165	178.5	830	30.0	24.1	0.4	90	4.4	(3)	Very high

*Number of dams more than 15 m high. Numbers in parentheses indicate new dams higher than 60 m under construction

Sources: Viviroli and others 2003, IUCN/WRI 2003, Nellemann 2005

qualitative judgments (Leemans 2000, Sala and others 2000). Species richness was found to be an insufficient indicator. On the one hand, it is hard to monitor the number of species in an area, but, more importantly, it may sometimes increase as original species are gradually replaced by new human-introduced invasives. Consequently the CBD has chosen a limited set of indicators to track this degradation process, selecting, among others, the "change in abundance of selected species" (CBD 2004). The GLOBIO 3.0 model was developed specifically to estimate this indicator. Biodiversity loss is here expressed as the percent age of original species abundance as found in undisturbed controls or in information about the diversity in the original state of the land use category in question (Nellemann 2005, Scholes and Biggs 2005).

Change in local biodiversity in the world's deserts, 1700–2050

In desert regions, examples of human impacts include fragmentation of wildlife habitats by roads and dams as illustrated above (Epps and others 2005), illegal exploitation of cactoids and reptiles (Goode and others 2005), poaching of wildlife (Spalton and others 1999, Combreau and others 2001, Mesochina and others 2003), as well as land degradation associated with human activity in oases, wadis, and sky islands (Zhao and others 2004). Introduction of new invasive species, like the African buffel grass (*Pennisetum ciliare*) introduced in Sonora to improve rangelands for cattle production, also includes major new threats to desert biodiversity (Franklin and others 2006). Currently, the desert biome holds an average abundance of original species of 68 per cent. Most of it is concentrated in hotspots or in transition zones between arid rangelands and true deserts. In 1700, mean abundance of original species was approximately 93 per cent (range 89–96%), dropping to 87 per cent in 1900 (range 83–91%). Given a proportional decline in abundance with either (a) population growth or (b) change in cropland, the rate of loss in original species abundance has been about 0.17 per cent per year (range 0.13–0.21 %) in the last century in deserts. This compares to the decline of 0.8–2.4 per cent of intact wilderness ecosystems per year. Declines are greatest in desert margins and mountainous

areas within deserts. Losses are also pronounced in coastal areas with high population density.

Future losses of biodiversity in deserts

Scenarios of change show that the rate of biodiversity loss in deserts may as much as double in the coming decades. These results are fairly similar compared to the global regime (CBD 2006). All four scenarios project a further decline in mean original species abundance from about 65 per cent in deserts in 2000 to a mean of 62.8 per cent by 2030 (range 60–65 %) and 58.3 per cent by 2050 (range 53–62 %; Figure 6.9).

Over a period of 50 years, the current global desert species abundance may thus drop by as much as 15 per cent — a dramatic decline given the relatively short timespan. The projected decline in biodiversity varies greatly among the scenarios. Remarkably, even a slowing of the rate of biodiversity loss to that of the mid 20th century would still mean a continued decline in the abundance of wildlife in desert regions. As for the scenarios in which the effect of land use and infrastructure development is modelled, the degree of decline varies greatly among and within the regions. Areas at desert biome boundaries or at higher elevations, such as in sky islands, are particularly prone and sensitive to change.

Ranking of pressures to biodiversity loss in deserts

Agriculture and human land use accounted for 41 per cent of the biodiversity loss by the year 2000. Fragmentation associated with infrastructure comes in at a close second (40 per cent). The relative share of the different factors varies among the scenarios, with climate change being the only one increasing in share for all four scenarios, from 6 per cent in 2000 to up to 14 per cent by 2050 (Thomas and others 2004), compared to a range of shares of 37–44 per cent for agriculture and 33–45 per cent for infrastructure. In deserts, infrastructure appears to play a major role in biodiversity losses, simply because it accelerates and facilitates human access to scattered and patchy hotspots of biodiversity where water is available, and because it increases fragmentation, which has been shown to have cascading adverse effects on ecosystems.

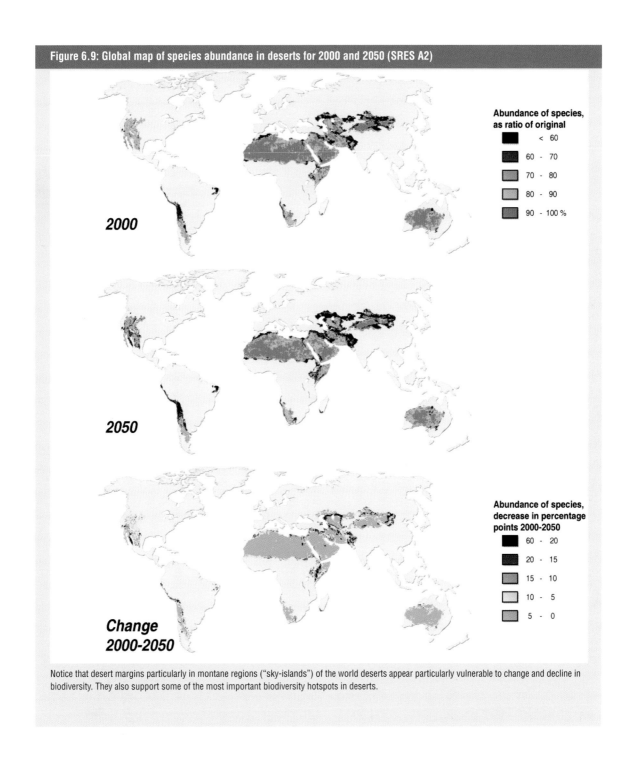

Figure 6.9: Global map of species abundance in deserts for 2000 and 2050 (SRES A2)

2000

Abundance of species, as ratio of original
■ < 60
■ 60 - 70
■ 70 - 80
□ 80 - 90
□ 90 - 100 %

2050

Change 2000-2050

Abundance of species, decrease in percentage points 2000-2050
■ 60 - 20
■ 20 - 15
□ 15 - 10
□ 10 - 5
□ 5 - 0

Notice that desert margins particularly in montane regions ("sky-islands") of the world deserts appear particularly vulnerable to change and decline in biodiversity. They also support some of the most important biodiversity hotspots in deserts.

COMMON TRENDS IN DESERT BIODIVERSITY CHANGE

Although there is some variation in what the different models suggest for the future, patterns nevertheless emerge which enable us to identify areas of particular concern. It appears that most of the dramatic changes take place at the desert margins. It also appears that wetlands, riparian zones and woodlands, particularly in and around sky-islands or desert montane areas, are targeted for development. The potential adverse impacts of development will be exacerbated by climate change in terms of temperature and precipitation. In concert, these factors can seriously compromise the function of desert montane areas as water towers, and thus the viability of species and human livelihoods that rely on them. There is an urgent need to increase their level and extent of protection.

Options for Action

The future of deserts, as natural and cultural landscapes, depends on our ability to develop their potential as providers of goods and services without degrading their ecological value in the face of increasing human pressures and possible climatic deterioration. Desired outcomes of action can be subsumed under two closely related concepts — human well-being and environmental sustainability.

The Millennium Ecosystem Assessment (MA) defined human well-being on the basis of five dimensions: the provision of the basic materials needed to sustain life, freedom and choice, health, good social relations, and personal security (MA 2003). Human well-being in deserts is generally below the global average. With the exception of the American Southwest and parts of the Arabian Peninsula, desert regions are characterized by comparatively high infant mortality rates and low economic performance, as expressed in their per capita gross domestic product (MA 2006). Not surprisingly, livelihood options in deserts are limited primarily by the scarcity of water, which, when coupled with poor infrastructure and social and political marginalization, negatively affects health and food security.

Environmental sustainability is a concept that emerged as part of the sustainable development discussion triggered by the World Commission on Environment and Development (1987). It implies that we should not deplete environmental goods and services, or at least that we should safeguard them (Goodland 1995). This must be compatible with qualitative improvements in human well-being rather than quantitative growth in production and consumption — particularly in deserts, given the heavy resource consumption of all desert development and the vulnerability of those resources.

The nature of human-environment interactions differs among and within societies in deserts, and consequently so do the options for action. The poor are particularly vulnerable to resource shortages or degradation, because their daily lives closely rely on their immediate environment. The better-off are more insulated from the variable environment and so in comparison hardly feel the effects of a drought. Importation of water and goods at affordable prices, as in Australia, the Arabian Peninsula or the American Southwest, can temporarily diminish the challenges of living in a desert, and diminish our awareness of limits and sustainability. Although they engage in the least sustainable rates of resource consumption, prosperous societies have (theoretically) the most options available to them to address environmental problems. However, there appear to be insufficient economic, social or ethical incentives to pursue a sustainable path, at least to this point.

In the face of globally increasing resource shortages, deserts hold a unique position with respect to the key resources of water and energy. Because water is in such short supply, deserts are the first environments to be forced to deal with water shortages. They should be in the forefront in developing and testing water-efficient technologies and policies, which are likely soon to become globally relevant as water demand increases worldwide. Energy might hold another opportunity for development in deserts, because of the implications of the increasing scarcity of fossil fuel resources and the impact their continued use might have on the global climate. More emphasis will have to be placed on renewable energy (solar, wind, geothermal) in our energy portfolio. The low cost of land and the abundance of solar energy should offer deserts an advantage on which they might capitalize.

RESOURCE MANAGEMENT FOR DESERT ECOSYSTEMS

The pulse-reserve character of desert ecosystems presents particular challenges to sustainable resource management. While the notion of sustainability implies some sort of balance between resource provision and extraction, the extreme variability inherent in desert ecosystems tends towards boom-and-bust cycles rather than a steady flow of environmental goods and services. As a result, sustainability is difficult to define for desert ecosystems and certainly cannot be achieved by prescribing a fixed carrying capacity (such as, the number of livestock that a particular region can sustain; Behnke and Scoones 1993).

Mobile, extensive forms of grazing have been found to be well adapted to the variable resource availability in desert ecosystems (Niamir-Fuller 1999). Traditional users have learned to exploit ecosystem cycles sustainably, through mobility and regulation as a means to moderate rangeland use (as in the collective reserve Hema system in Arabia; see Chapter 2). In contrast, modern trends toward the sedentarization of pastoralists and the provision of subsidized supplemental animal feed, though implemented in the interest of economic sustainability, increase pressures on ecosystems by allowing for long periods of stay (Al-Rowaily 1999).

Sustainable resource management policies must respond to the pulse-reserve character of the desert ecosystem by supporting mobile or otherwise flexible systems, which can respond to the variable and unpredictable desert environment and still remain economically viable over long periods of time. This support can take the form of providing mobile services (medical care, schooling), encouraging risk spreading through common property management (Hesse and Trench 2000), and providing timely and accurate information about the state of pastures.

Mitigating the "bust" part of the cycle is another important component of the sustainable management of desert ecosystems. This includes not only emergency support during drought crises, but also proactive management to increase human and societal resilience, by creating diverse rural income opportunities, providing support for animal marketing, providing credit, and establishing other forms of insurance that can sustain rural livelihoods during times of stress (Box 6.4). An alternative is to encourage urbanization that essentially removes pressures on rural natural resources (Portnov and Safriel 2004).

MAKING USE OF MODERN TECHNOLOGY
Traditional wisdom on coping with drought (Mortimore 1998), complemented by cutting-edge science and information technology (for example, drought forecasting and climate change scenarios), holds great potential for sustainable resource management. If we can have better information about the near future, we can plan better how to deal with it.

Although drought alone cannot be held responsible for causing food insecurity (Sen 1981), some

Box 6.4: An approach to risk management in deserts

An unusual approach to risk management in the deserts of Namibia has grown out of two systems used in the deserts and neighbouring semi-arid environments. The systems are closely related and involve institutional development and capacity-building of traditional desert residents, who have found themselves marginalized in a newly-independent country. Community-based natural resource management, with its focus on communal conservancies to protect wildlife and ensure its sustainable use and management, has proven to be a promising approach (NACSO 2004). At least eight of 29 registered communities are situated in the Namib Desert, while another ten border it closely. Through the robust institutional structures established by communities to manage and use natural resources in profitable and sustainable ways, their livelihoods are improved through the use of high-value desert wildlife for tourism, trophy hunting and meat. This approach provides the desert communities with alternative incomes when drought, or other risks, occur.

Rural Namibians holding a community-scale FIRM meeting.
Source: Desert Research Foundation of Namibia

Somewhat similar, but more broadly focused, are the Forums for Integrated Resource Management (FIRM) through which desert inhabitants take the lead in coordinating their own development trajectories (Kambatuku 2003). By establishing a FIRM, and inviting all relevant service providers (including agricultural and water extension officers, education and health authorities, conservancy representatives, regional councillors and traditional leaders) to participate in coordinated development planning and regular monitoring of progress, desert peoples can take charge of their own development according to their own preferences, and plan for and manage anticipated risks as they occur.

Box author: Mary Seely

desert regions face food insecurity and increases in excess mortality during prolonged drought periods. USAID has initiated a famine early warning system project (FEWS NET: http://www.fews.net/), which provides drought early warning and vulnerability information for drought-menaced African countries, both in semi-arid drylands and in deserts.

The objectives of interventions triggered by the early-warning information range from saving lives in response to immediate emergencies (for example, emergency food programs and livestock health interventions), to saving both lives and livelihoods by reducing exposure to risk and developing diverse opportunities to generate income (such as, crafts and other off-farm employment) other than from agriculture.

While the activities of FEWS NET tend to focus on the most vulnerable human populations, an example from a more robust setting is the Rangeview project (http://rangeview.arizona.edu) that sprang from an initiative to make geospatial technology accessible to a range of users. Launched in 2000 to provide rangeland managers in the American Southwest with satellite-derived information about the status and trends of vegetation greenness, it evolved into a decision-support tool and has meanwhile been extended to cover the entire United States. Similar initiatives, making use of the Internet as an inexpensive means of information exchange, could be beneficial to rangeland management in desert regions around the world.

This type of tool helps natural resource managers to understand what is happening now, and compare it with what has been happening over time and come to an appreciation of where we are at any point in time. As our ability to understand and model the global climate system improves, we are able to develop increasingly useful seasonal weather forecasts for large regions and long-term scenarios. These can be used to plan for adapting to climate variability and change, which are expected to be increasingly inevitable during the next decades (Dessai and others 2005).

Technical knowledge and reliable forecasts alone, however, are insufficient; they need to be implemented to the benefit of people under a given set of circumstances. Climate change adaptation planning therefore must include the identification of vulnerable population groups and the exploration of effective and affordable livelihood strategies during times of climatic stress. Perhaps most importantly, there need to be systems in place that have the political will and institutional capacity to act on the most likely scenarios, or at least to accommodate some outcomes.

RENEWABLE ENERGY FROM THE DESERT

The provision of clean and affordable energy is one of the most critical problems that confronts human development. The current world energy system is dominated by fossil fuels and will fall short in meeting the energy demands of a projected world population of 10 billion by the middle of the century (Smalley 2005). In prosperous nations, energy conservation through improved efficiency offers one possibility to reduce demand. In much of the developing world, however, conservation is meaningless because little energy is currently used and their total demand can only increase as they develop. Although currently not yet profitable, renewable energy resources could account for one-third to one-half of the global energy supply by 2050, based on price competition (Shell International 2001).

Their continuously high solar radiation makes deserts ideal locations for both small decentralized units and large solar cell installations, the potential reach of which is not limited to deserts (see Chapter 5). Apart from technological feasibility, the adoption of solar energy as an alternative to fossil fuels depends on the global as well as national policy environments and concrete implementation strategies. Possible incentives to encourage the shift towards renewable energy sources include taxes on pollution-generating practices, such as the burning of fossil fuels, while providing loans, grants or subsidies for the use of solar and other renewable energy resources. In addition, allocating funds for relevant technological research and training and for marketing of solar technologies, and raising public awareness of renewable resources as a clean alternative to fossil fuels, could be used to help promote the use of solar energy.

THE "SOFT PATH" FOR WATER DEVELOPMENT

Twentieth-century water policies were largely dominated by the construction of massive water extraction, storage and transport infrastructure, which brought benefits, such as the expansion of irrigated agriculture in deserts, but at substantial environmental costs (Gleick 2003), as illustrated by the examples of the Aral Basin in Central Asia and the Imperial Valley in the southwestern United States (see Chapter 5).

Analogous to the "soft path" for energy policy advocated by Lovins (1976), leading water experts strongly promote a "soft path" for water development in response to the impending global water problems (Gleick 2003; Rijsberman 2004). This soft path should focus on water-use efficiency and the control of demand rather than building ever bigger dams and endlessly developing new sources of water. Deserts, as the first environments confronted with water shortages and forced to rethink water use priorities, should be among the forerunners in developing and testing innovative, and globally relevant, technologies and policies.

The implementation of water conservation measures and improvements in water-use efficiencies needs to be supported by economic and institutional structures. In many desert regions, water prices currently do not reflect the value of water. A strategy to discourage wasteful water consumption, which at the same time contributes to more equitable access to water, is to support low-income and low-volume users with transparent subsidies, financed by excessive water consumers. Raising public awareness about the need to conserve water is particularly important for new migrants into deserts who have not developed a "sense of place", such as those moving into the desert cities of the American Southwest.

Small-scale decentralized water supply facilities and the involvement of communities in the decision-making process about water management, allocation and use ensure more equitable access to water and potentially lower environmental impacts than the massive centrally-planned water schemes of the 20th century (Gleick 2003). In the communal parts of Namibia, for example, water point committees have been set up as part of a larger decentralization policy, which are responsible for the provision of water to the community and the maintenance of communal water installations (Werner 2000).

Promotion of only high value-added uses of water can improve water efficiency: for example, the high-tech industrial sector enhances the value of each cubic metre of water used many times more than the agricultural sector (Gleick 2001). Within the agricultural sector, one possibility to improve water efficiency is to restrict irrigated agriculture in deserts to high-value crops (for example, dates) or aquaculture (see Chapter 3), whereas lower value crops (for example, maize) can be imported from regions better endowed with water. Despite the risks involved with abandoning their food independence, many water-poor countries already choose to import food rather than growing it, creating a virtual flow of water, which is contained in the imported food products or other commodities that require high water inputs. With increasing globalization, the import of "virtual water" becomes a tool in water-resource management, which can be used to relieve pressure on scarce water resources in deserts.

Concluding Remarks

What will the future hold for the deserts of the world? Based on the foregoing analyses, we can assert with some confidence a number of simple predictions: deserts are and will remain constrained in their productive potential by their very nature — the widely varying conditions of the desert ecosystem, the scarcity of water, and the oscillating resource variability. This basic nature of deserts will not change in the forseeable future. However, global climate change, coupled with increased population pressure, particularly in the desert margins, montane areas and wetlands, is likely to affect the more productive desert areas and pose some new and significant threats to biodiversity and sensitive endemic species. By contrast, the hyper-arid cores of the desert biome — the vast wilderness of the "deep" desert — are going to be less affected by mounting pressures on their fragile biological resources. Water depletion, on the other hand, as well as salinization of irrigated agricultural soils, are likely to continue as two of the main environmental problems in many deserts, encouraged by modern technologies for groundwater prospecting and pumping. A sort of modern itinerant agriculture has already emerged, where large barren areas of salinized agricultural soils are left behind as groundwater resources become exhausted and agricultural operations move on to new lands and to new untapped aquifers.

These predictions, however, are not fixed and unchangeable. The scenario analyses discussed in the previous sections show that there is a wide range of possible outcomes for deserts, an array of alternative futures. Whether deserts will follow a path of intensive development, industrial-scale agriculture projects and mega-cities attracting massive immigration — a vision that has been called, somewhat sarcastically, the "Cadillac Desert" (Reisner 1986) — or an alternative path of sustainable development, spurred by a "sense of place" and prioritizing the desert environment and the traditional culture of local communities, will be largely determined by our common visions and collective action taken to fulfil them.

In reality, current development in many deserts seems to suffer from a lack of vision altogether. Few, if any, coordinated programs exist for either development or conservation of the land. The unique values and limitations of the desert are rarely acknowledged. Development schemes, such as programmes for irrigated agriculture or mass tourism, tend to spring up haphazardly with no attempt to coordinate them or to plan for their long-term sustainability. Immigration to the desert is often random and opportunity-driven, and new settlements sprawl over valuable landscapes and create problems for water supply and waste management. Without proper planning and a vision of sustainability, traditional lifestyles atrophy and indigenous knowledge is lost, victims of short-term, ephemeral economic projects.

So what is to be done; what vision should be pursued for the successful long-term development of deserts, especially in developing countries? Quite clearly, a continuation of the energy- and water-intensive development model will lead to even more severe water depletion and degradation than is observed today, followed by potential conflict over water resources among users, escalating costs of supply, and the continuation of a non-renewable model in which water, often under immense subsidies, is, paradoxically, used for low-value purposes. At the other extreme, increased isolationism with exclusive reliance on traditional knowledge runs the risk of losing access to new sustainable technologies and might lead to diminished opportunities for younger generations, and, eventually, to reduced livelihood and economic development options.

A new, more balanced vision is needed, where deserts and their inhabitants are valued both by governments and civil society; where sustainability and the well-being of desert people are given highest priority; where desert development is guided by a long planning horizon and based on an acute understanding of the limitations and potential of these very unique environments; where market forces are harnessed to promote a desert-compatible development, such as low-impact services or high-technology development; where traditional livelihoods are given the opportunity to survive with dignity; and where wetlands, oases, desert mountains and other environments at risk are protected.

Decisions can and should be made not to change the desert, but to live with it and preserve its resources for the future. The active participation of community groups in each desert for the development of a common vision is a fundamental condition for the successful formulation and implementation of policies towards an environmentally sustainable future. Desert peoples can, and should, take charge of their own development, plan for risks, and adapt to changing conditions while preserving their deep connections to these remarkable landscapes. The challenge remains to harness not only local, but also global policy mechanisms and market incentives to develop a viable future for our deserts, where both the successful protection of the environment and economic development opportunities are achieved.

REFERENCES

Ahlenius, H., Johnsen, K., and Nellemann, C. (2005). *Vital Graphics — People and global heritage on our last wild shores*. United Nations Environment Programme, Arendal (available at http://www.grida.no)

Alcamo, J., Kreileman, G.J.J., Bollen, J.C., van den Born, G.J., Gerlagh, R., Krol, M.S., Toet, A.M.C., and de Vries, H.J.M. (1996). Baseline scenarios of global environmental change. *Global Environmental Change* 6: 261–303

Al-Rowaily, S.L.R. (1999). Rangeland of Saudi Arabia and the "Tragedy of Commons". *Rangelands* 21: 27–29

Alward, R.D., Detling, J.K., and Milchunas, D.G. (1999). Grassland vegetation changes and nocturnal global warming. *Science* 283: 229–231

Andrews, A. (1990). Fragmentation of habitat by roads and utility corridors: A review. *Australian Zoologist* 26: 130–141

ArabNews (2005). "Riyadh Population Crosses 4 Million Mark." March 7, 2005. http://www.arabnews.com/ [Accessed 19 April 2006]

Archer, D.R., and Fowler, H.J. (2004). Spatial and temporal variations in precipitation in the Upper Indus Basin, global teleconnections and hydrological implications. *Hydrology and Earth System Sciences 8: 47–61*

Arnell, N.W. (1999). Climate change and global water resources. *Global Environmental Change* 9: 831–849

Barnett, T.P., Adam, J.C., and Lettenmaier, D.P. (2005). Potential impacts of a warming climate on water availability in snow-dominated regions. *Nature* 438: 303–309

Behnke, R.H., and Scoones, I. (1993). Rethinking Range Ecology: Implications for Rangeland Management in Africa. In *Range Ecology at Non-equilibrium. New Models of Natural Variability and Pastoral Adaptation in African Savannas* (eds. R.H. Behnke, I. Scoones, and C. Kerven) pp. 1–30. Overseas Development Institute, London

Bekenov, A.B., Grachev, I.A., and Milner-Gulland, E.J. (1998). The ecology and management of the Saiga antelope in Kazakhstan. *Mammal Review* 28: 1-52

Bleich, V.C. Wehausen, J.D., and Holl, S.A. (1990). Desert-dwelling mountain sheep — conservation implications of a naturally fragmented population. *Conservation Biology* 4: 383–390

Blyth, S., Groombridge, B., Lysenko, I., Miles, L., and Newton, A. (2002). *Mountain watch — environmental change and sustainable development in mountains*. United Nations Environment Programme, UNEP- World Conservation Monitoring Centre, Cambridge

Böhner, J., and Lehmkuhl, F. (2005). Environmental change modelling for Central and High Asia: Pleistocene, present and future scenarios. *Boreas* 34: 220–231

Brown, K. (2000). Ghost Towns Tell Tales of Ecological Boom and Bust (Mojave Desert Study). *Science* 290: 35–37

Butler, D. (2003). Facing up to endangered apes. *Nature* 426: 369

Caldecott, J., and Miles, L. (2005). *World Atlas of Great Apes and their conservation*. United Nations Environment Programme, UNEP-World Conservation Monitoring Centre, Cambridge

CBD (2002). *Report of the sixth meeting of the conference of the parties to the Convention on Biological Diversity*. Decision VI/22: Forest Biological Diversity. UNEP/CBD/COP/6/20. United Nations Environment Programme, Montreal

CBD (2004). Indicators for assessing progress towards, and communicating, the 2010 target at the global level. UNEP/CBD/SBSTTA/10/9. United Nations Environment Programme, Montreal

CBD (2006). *Global Biodiversity Outlook 2*. Secretariat of the Convention on Biological Diversity, Montreal, Canada (available online at http://www.biodiv.org/GBO2)

CIESIN (2000). Gridded Population of the World (GPW), Version 2. Center for International Earth Science Information Network (CIESIN), Columbia University, Palisades, NY. http://sedac.ciesin.columbia.edu/plue/gpw [Accessed 19 April 2006]

Claussen, M., Brovkin, V., Ganopoliski, A., Kubatzki, C. and Petoukhov, V. (2003). Climate Change in Northern Africa: The Past is not the Future. *Climatic Change* 57: 99-118

Cohen, J.E. (2003). Human Population: The Next Half Century. *Science* 302: 1172–1175

Combreau, O., Launay, F., and Lawrence, M. (2001). An assessment of annual mortality rates in adult-sized migrant houbara bustards (Chlamydotis [undulata] macqueenii). *Animal Conservation* 4: 133–141

Corsale, A. (2005): *Popolazione e ambiente nelle oasi della fascia predesertica del Maghreb: evoluzione socio-economica in relazione alle risorse idriche* (English: People and Environment in the oases of Morocco). Doctoral thesis, University of Cagliari, Italy

Dessai, S., Lu, X., and Risbey, J.S. (2005). On the role of climate scenarios for adaptation planning. *Global Environmental Change* 15: 87–97

Di Castri, F. (2000). Ecology in a Context of Economic Globalization. *BioScience* 50: 321–332

Edwards, T., Schwalbe, C.R., Swann, D.E., and Goldberg, C.S. (2004). Implications of anthropogenic landscape change on inter-population movements of the desert tortoise (Gopherus agassizii). *Conservation Genetics* 5: 485–499

Epps, C. W., Palsboll, P. J., Wehausen, J. D., Roderick, G. K., Ramey, R. R., and McCullough, D. R. (2005). Highways block gene flow and cause a rapid decline in genetic diversity of desert bighorn sheep. *Ecology letters* 10: 1029-1038

Ezcurra, E., Valiente-Banuet, A., Flores-Villela, O., and Vazquez, E.. (2001). Vulnerability to global environmental change in natural systems and rural areas: A question of latitude? In *Global environmental risk* (eds. J.X. Kasperson and R.E. Kasperson) pp. 217–246. United Nations University Press, Tokyo

Falkenmark, M., and Widstrand, C. (1992). Population and Water Resources: A Delicate Balance. *Population Bulletin* 47. Population Reference Bureau, Washington, DC

Fekete, B.M., Vorosmarty, C.J., and Grabs, W. (1999). *Global, Composite Runoff Fields Based on Observed River Discharge and Simulated Water Balances*, GRDC Report 22. Global Runoff Data Center, Koblenz, Germany

Forman, R.T.T., and Alexander, L.E. (1998). Roads and their major ecological effects. *Annual Review of Ecology and Systematics* 29: 207–231

Franklin, K.A., Lyons, K., Nagler, P.L., Lampkin, D., Glenn, E.P., Molina-Freaner, F., Markow, T., and Huete, A.R. (2006). Buffelgrass (Pennisetum ciliare) land conversion and productivity in the plains of Sonora, Mexico. *Biological Conservation* 127: 62–71

Garba, S.B. (2004). Managing urban growth and development in the Riyadh metropolitan area, Saudi Arabia. *Habitat International* 28: 593–608

Glantz, M.H. (ed.) (1994). *Drought Follows the Plow. Cultivating Marginal Areas*. Cambridge University Press, Cambridge

Gleick, P.H. (2001). Making every drop count. *Scientific American Magazine* 284: 40–46

Gleick, P.H. (2003). Global Freshwater Resources: Soft-Path Solutions for the 21st Century. *Science* 302: 1524–1528

Gleick, P.H. (ed.) (2006). *The World's Water 2006–2007. The Biennial Report on Freshwater Resources*. Island Press, Washington, DC

Goldewijk, K. (2001). Estimating global land use change over the past 300 years: the HYDE database. *Global Biogeochemical Cycles* 15: 417–433

Goode, M.J., Horrace, W.C., Sredl, M.J., and Howland, J. (2005). Habitat destruction by collectors associated with decreased abundance of rock-dwelling lizards. *Biological Conservation* 125: 47–54

Goodland, R. (1995). The concept of environmental sustainability. *Annual Review of Ecology and Systematics* 26: 1–24

Gutzwiller, K.J., and Barrow, K.C. (2003). Influences of roads and development on bird communities in protected Chihuahan Desert landscapes. *Biological Conservation* 113: 223–237

Haarsma, R.J., Selten, F.M., Weber, S.L., and Kliphuis, M. (2005). Sahel rainfall variability and response to greenhouse warming. *Geophysical Research Letters* 32: L17702, doi:10.1029/2005GL023232

Held, I.M., Delworth, T.L., Lu, J., Findell, K.L., and Knutson, T.R. (2005). Simulation of Sahel drought in the 20th and 21st centuries.

Appendix 1

Area and type of the desert ecoregions of the world

Realm names are shown in bold. Ecoregion names are modified from WWF Terrestrial Ecoregions of the World (Olson and others 2001), and further characterized by desert type: Lowland or Plain Desert (LD), Coastal Desert (CD) or Montane Relict (MR).

Ecoregion name	Type	WWF code[A]	Total area (km^2 × 10^3)[A]	Area converted (ha × 10^3)[B]	Area protected (IUCN I-IV) (ha × 10^3)[B]	Area protected (all IUCN) (ha × 10^3)[B]
Afrotropic						
Al Hajar montane woodlands	MR	AT0801	25.49	5	-	-
Arabian Peninsula coastal fog desert	CD	AT1302	82.69	163	1,529	1,529
Eritrean coastal desert	CD	AT1304	4.58	1	-	-
Ethiopian montane forests [†]	MR	AT0112	0.57	14,570	808	2,627
Ethiopian xeric grassland and shrublands	LD	AT1305	152.52	273	686	686
Gulf of Oman desert and semi-desert	CD	AT1306	62.30	29	18	91
Madagascar spiny thickets	CD	AT1311	43.29	75	62	62
Nama Karoo	LD	AT1314	350.73	695	730	765
Namib desert	CD	AT1315	80.69	3	4,629	6,287
Northern Namib (Skeleton Coast or Kaokoveld desert)	CD	AT1310	45.59	1	2,099	2,925
Somali Acacia-Commiphora bushlands and thickets	LD	AT0715	1049.30	9,997	2,898	11,863
Somali montane xeric woodlands	MR	AT1319	62.38	22	-	-
Southwestern Arabian foothills savanna	LD	AT1320	273.76	2,057	-	58
Southwestern Arabian montane woodlands	MR	AT1321	86.63	719	-	-
Succulent Karoo	CD	AT1322	102.59	26	190	257
Australasia						
Carnarvon xeric shrublands	LD	AA1301	90.29	576	605	605
Central Ranges xeric scrub	LD	AA1302	281.05	1,950	331	5,271
Gibson desert	LD	AA1303	155.52	-	1,853	5,136
Great Sandy-Tanami desert	LD	AA1304	820.77	909	1,629	2,523
Great Victoria desert	LD	AA1305	423.73	157	5,437	6,217
Nullarbor Plains xeric shrublands	LD	AA1306	195.00	21	2,420	2,729
Pilbara shrublands	LD	AA1307	179.31	45	1,139	1,140
Simpson desert	LD	AA1308	583.39	694	2,052	5,424
Tirari-Stuart stony desert	LD	AA1309	376.49	747	1,393	1,795
Western Australian Mulga shrublands	LD	AA1310	459.62	67	513	521
Indo-Malay						
Indus Valley desert	LD	IM1302	19.48	1,201	-	1,220
Thar desert	LD	IM1304	238.25	9,423	2,514	4,152
Nearctic						
Baja California desert	CD	NA1301	77.59	1	348	3,972
Chihuahuan desert	LD	NA1303	508.89	515	480	3,389
Eastern Madrean Archipelago [†]	MR	NA0303	20.98	759	227	601
Great Basin montane forests	MR	NA0515	5.79	-	170	487
Great Basin shrub steppe	LD	NA1305	336.21	1,327	580	7,477
Gulf of California xeric scrub	CD	NA1306	23.54	2	40	847
Meseta Central matorral	LD	NA1307	124.98	255	12	140
Mojave desert	LD	NA1308	130.65	183	1,715	7,025
Sierra Juárez and San Pedro Mártir pine-oak forests	MR	NA0526	4.00	-	67	68
Sonoran desert	LD	NA1310	222.84	1,128	1,800	6,674
Western Madrean Archipelago [†]	MR	NA0302	16.28	714	165	1,259
Neotropic						
Atacama desert	CD	NT1303	104.90	24	122	122
Central Andean dry puna	LD	NT1001	254.93	407	1,626	3,615
High Monte	LD	NT1010	116.57	22	377	1,395
Low Monte	LD	NT0802	353.64	163	742	1,949
Sechura desert	CD	NT1315	184.21	757	22	369
Palearctic						
Afghan Mountains semi-desert	MR	PA1301	13.68	179	54	54
Alashan Plateau semi-desert	LD	PA1302	674.35	1,281	8,093	12,342
Arabian desert and East Sahero-Arabian xeric shrublands	LD	PA1303	1847.46	548	4,315	62,725
Atlantic coastal desert	CD	PA1304	39.89	13	523	706
Azerbaijan shrub desert and steppe	LD	PA1305	64.09	1,543	297	367
Badghyz and Karabil semi-desert	LD	PA1306	133.65	3,020	306	327

Ecoregion name	Type	WWF code[A]	Total area (km² × 10³)[A]	Area converted (ha × 10³)[B]	Area protected (IUCN I-IV) (ha × 10³)[B]	Area protected (all IUCN) (ha × 10³)[B]
Caspian lowland desert	LD	PA1308	267.98	1,516	1,614	1,627
Central Asian northern desert	LD	PA1310	663.85	2,823	1,029	1,032
Central Asian southern desert	LD	PA1312	567.38	703	627	627
Central Persian desert basins	LD	PA1313	580.66	394	3,047	5,203
Eastern Gobi desert steppe	LD	PA1314	202.37	207	074	1,000
Gobi Lakes Valley desert steppe	LD	PA1315	139.71	742	220	220
Great Lakes Basin desert steppe	LD	PA1316	157.71	1,223	2,054	2,067
Gulf desert and semi-desert	CD	PA1323	72.52	16	8	733
Junggar Basin semi-desert	LD	PA1317	304.94	2,903	1,165	3,342
Kuh Rud and Eastern Iran montane woodlands	MR	PA1009	126.22	99	76	271
Mesopotamian shrub desert	LD	PA1320	210.91	165	81	188
North Saharan steppe and woodlands	LD	PA1321	1673.78	412	76	1,533
Qaidam Basin semi-desert	LD	PA1324	192.15	87	-	1,975
Red Sea coastal desert	CD	PA1333	59.20	9	968	1,397
Red Sea Nubo-Sindian tropical desert and semi-desert	LD	PA1325	649.66	28	1,833	16,263
Registan-North Pakistan sandy desert	LD	PA1326	277.00	2,053	139	441
Sahara desert	LD	PA1327	4629.42	603	7,137	10,439
South Iran Nubo-Sindian desert and semi-desert	LD	PA1328	350.89	304	952	1,956
South Saharan steppe and woodlands	LD	PA1329	1098.25	104	6,822	7,020
Taklimakan desert	LD	PA1330	742.66	3,321	-	8,942
Tarim Basin deciduous forests and steppe	MR	PA0442	54.53	145	-	408
Tibesti-Jebel Uweinat montane xeric woodlands	MR	PA1331	82.01	-	-	-
West Saharan montane xeric woodlands [†]	MR	PA1332	177.28	4	10,734	10,734

[†] Only the geographic area of the WWF ecoregion that lies within the desert biome as defined in this report is presented in the "area" column. The land conversion and land protection statistics also presented in the table, however, represent those for the whole ecoregion, not just the subset of the ecoregion that occurs within the desert biome.

(-) Insufficient data available.

Data compiled or derived by The Nature Conservancy from the following sources:

[A] Olson and others (2001). BioScience 51: 933–938.

[B] Hoekstra and others (2005). Ecology Letters 8:23–29.

Appendix 2

Species richness, human population, and human footprint of the desert ecoregions of the world

Ecoregion name	Plant species richness[c]	Vertebrate species richness[c]	Endemic vertebrates species[c]	Threatened vertebrate species[D]	Human population density (people/km²)[E]	Human Footprint[F]
Afrotropic						
Al Hajar montane woodlands	700	113	4	6	19.2	23.7
Arabian Peninsula coastal fog desert	1000	372	8	11	35.2	22.0
Eritrean coastal desert	100	282	-	10	6.6	24.5
Ethiopian montane forests [†]	4000	703	17	18	76.1	30.5
Ethiopian xeric grassland and shrublands	850	520	4	19	17.6	20.6
Gulf of Oman desert and semi-desert	300	344	1	8	45.7	25.3
Madagascar spiny thickets	1100	288	38	19	21.9	24.7
Nama Karoo	1100	481	6	18	1.8	11.1
Namib desert	1000	301	7	8	1.2	4.8
Northern Namib (Skeleton Coast or Kaokoveld desert)	500	360	8	5	3.2	10.4
Somali Acacia-Commiphora bushlands and thickets	2600	904	51	29	13.8	18.9
Somali montane xeric woodlands	1500	277	7	13	5.0	13.3
Southwestern Arabian foothills savanna	400	350	7	13	46.1	21.8
Southwestern Arabian montane woodlands	2000	200	6	10	91.0	24.6
Succulent Karoo	4850	409	20	21	3.2	11.4
Australasia						
Carnarvon xeric shrublands	700	425	16	9	0.2	8.7
Central Ranges xeric scrub	1400	376	5	9	0.1	4.7
Gibson desert	400	262	-	6	0.0	1.2
Great Sandy-Tanami desert	500	404	4	11	0.0	1.1
Great Victoria desert	400	343	2	11	0.0	2.3
Nullarbor Plains xeric shrublands	200	301	-	8	0.1	5.7
Pilbara shrublands	800	389	15	7	0.2	8.1
Simpson desert	800	403	3	11	0.0	1.9
Tirari-Stuart stony desert	800	464	3	12	0.2	6.2
Western Australian Mulga shrublands	900	374	3	12	0.0	6.0
Indo-Malay						
Indus Valley desert	400	240	-	5	200.2	36.0
Thar desert	650	240	-	10	102.5	30.1
Nearctic						
Baja California desert	3500	658	10	5	6.9	17.6
Chihuahuan desert	3200	693	24	23	331.8	36.2
Eastern Madrean Archipelago [†]	3300	859	15	13	35.3	26.5
Great Basin montane forests	1109	358	2	3	8.3	19.6
Great Basin shrub steppe	2556	489	-	7	2.8	12.8
Gulf of California xeric scrub	2519	451	1	9	6.8	16.5
Meseta Central matorral	2519	460	2	10	8.2	11.6
Mojave desert	2490	498	3	12	17.7	15.0
Sierra Juárez and San Pedro Mártir pine-oak forests	1729	440	1	3	15.2	28.1
Sonoran desert	3500	656	3	16	44.2	30.0
Western Madrean Archipelago [†]	4400	840	33	16	10.1	20.6
Neotropic						
Atacama desert	800	120	3	6	5.9	16.2
Central Andean dry puna	1100	320	16	10	3.6	12.9
High Monte	-	349	12	9	3.4	12.8
Low Monte	700	307	5	10	7.8	14.9
Sechura desert	1050	320	29	12	67.8	23.6
Palearctic						
Afghan Mountains semi-desert	600	95	-	7	17.5	28.1
Alashan Plateau semi-desert	700	559	3	24	9.9	9.5
Arabian desert and East Sahero-Arabian xeric shrublands	900	508	6	26	14.9	15.7
Atlantic coastal desert	300	122	-	5	7.3	10.6
Azerbaijan shrub desert and steppe	1500	396	1	17	120.0	36.9
Badghyz and Karabil semi-desert	1100	343	2	11	36.8	28.6
Caspian lowland desert	700	392	1	18	8.1	14.2
Central Asian northern desert	800	353	-	20	4.5	12.4

Ecoregion name	Plant species richness[c]	Vertebrate species richness[c]	Endemic vertebrates species[c]	Threatened vertebrate species[D]	Human population density (people/km²)[E]	Human Footprint[F]
Central Asian southern desert	900	305	-	15	11.5	17.4
Central Persian desert basins	1300	318	6	18	34.9	19.9
Eastern Gobi desert steppe	400	318	-	13	3.3	11.3
Gobi Lakes Valley desert steppe	400	220	-	16	1.0	10.1
Great Lakes Basin desert steppe	600	292	-	27	1.3	10.4
Gulf desert and semi-desert	350	298	-	4	40.4	24.5
Junggar Basin semi-desert	1200	352	-	21	16.6	15.0
Kuh Rud and Eastern Iran montane woodlands	1500	257	-	12	17.3	25.1
Mesopotamian shrub desert	1100	422	2	16	29.3	26.8
North Saharan steppe and woodlands	1150	330	6	15	5.9	7.2
Qaidam Basin semi-desert	250	207	-	12	1.2	6.8
Red Sea coastal desert	350	154	1	-	5.0	13.9
Red Sea Nubo-Sindian tropical desert and semi-desert	900	433	3	12	7.9	13.8
Registan-North Pakistan sandy desert	700	393	7	18	11.2	17.8
Sahara desert	500	335	3	16	2.0	2.6
South Iran Nubo-Sindian desert and semi-desert	900	483	5	24	20.7	26.0
South Saharan steppe and woodlands	500	348	1	18	1.1	3.8
Taklimakan desert	400	409	1	22	10.8	11.7
Tarim Basin deciduous forests and steppe	600	241	-	8	8.6	13.3
Tibesti-Jebel Uweinat montane xeric woodlands	580	106	1	9	0.2	0.5
West Saharan montane xeric woodlands [†]	550	138	-	8	0.3	2.2

[†] Only a portion of the WWF ecoregion (codes provided in Appendix 1) occurs within the desert biome as defined in this report. The statistics presented in the table, however, are for the whole ecoregion, and not just the subset within the desert biome.
(-) Insufficient data available.

Data compiled or derived by The Nature Conservancy from the following sources:
[c] *Lamoreux and others (2006). Nature 440: 212–214. (Data available at http://www.worldwildlife.org/wildfinder).*
[D] *Loucks and others (in review). World Wildlife Fund, Washington D.C. [Based on IUCN Red List].*
[E] *CIESIN Columbia University, IFPRI, the World Bank, and CIAT (2004). Global Rural-Urban Mapping Project (GRUMP). (http://sedac.ciesin.columbia.edu/gpw).*
[F] *Sanderson and others (2002). Bioscience 52: 891–904 (http://www.ciesin.columbia.edu/wild_areas/) [Ecoregion averages scaled from 0 -100].*

List of Authors, Reviewers and Contributors

Editor:

Exequiel Ezcurra, San Diego Natural History Museum, San Diego, California, USA

Technical copyeditors:

Margaret Dykens, San Diego Natural History Museum, San Diego, California, USA

Isolina Martínez, San Diego Natural History Museum, San Diego, California, USA

EXECUTIVE SUMMARY

Authors:

Timo Maukonen, UNEP DEWA, Nairobi, Kenya

Kakuko Nagatani, DEWA, UNEP-ROLAC, Mexico City, Mexico

Marcus Lee, UNEP DEWA, Nairobi, Kenya

Exequiel Ezcurra, San Diego Natural History Museum, San Diego, California, USA

CHAPTER 1: NATURAL HISTORY AND EVOLUTION OF THE WORLD'S DESERTS

Lead author:

Exequiel Ezcurra, San Diego Natural History Museum, San Diego, California, USA

Contributing authors:

Charlotte González, San Diego Natural History Museum, San Diego, California, USA

Eric Mellink, Centro de Investigación Científica y Educación Superior de Ensenada (CICESE), Ensenada, Mexico

Scott Morrison, The Nature Conservancy, California, USA

Andrew Warren, Department of Geography, University College London, UK

Elisabet Wehncke, San Diego Natural History Museum, San Diego, California, USA

Box authors:

David Dent, ISRIC - World Soil Information, Wageningen, The Netherlands

Paul Driessen, ITC, Enschede, The Netherlands

CHAPTER 2 - PEOPLE AND DESERTS

Lead author:

Mary Seely, Desert Research Foundation of Namibia, Windhoek, Namibia

Contributing authors:

Ahmed I. Al-Amoud, College of Agriculture, Riyadh, Saudia Arabia

Dawn Chatty, University of Oxford, UK

Joh Henschel, Gobabeb Training and Research Centre, Gobabeb, Namibia

Jill Kinahan, Quaternary Research Services, Windhoek, Namibia

John Kinahan, Quaternary Research Services, Windhoek, Namibia

Patrik Klintenberg, Desert Research Foundation of Namibia, Windhoek, Namibia

Alejandro León, Las Cardas Experiment Station, Chile

Scott Morrison, The Nature Conservancy, San Diego, California, USA

Conrad Roedern, Solar Age, Windhoek, Namibia

Box authors:

Elena Abraham, Instituto Argentino de Investigaciones de las Zonas Áridas, Mendoza, Argentina

Richard S. Felger, Drylands Institute, Tucson, Arizona, USA

Pietro Laureano, IPOGEA — Centro Studi Italiano sulle Conoscenze Tradicionali e Locali, Matera, Italy

David Mouat, Desert Research Institute, Reno, Nevada

Boris A. Portnov, University of Haifa, Haifa, Israel

Uriel Safriel, The Hebrew University of Jerusalem, Jerusalem, Israel

Sabine Schmidt, Initiative for People Centred Conservation, New Zealand Nature Institute, Ulan Bator, Mongolia

Maurizio Sciortino, Centro Ricerche Casaccia — ENEA, Rome, Italy

Andrew Warren, University College London, London, UK

Daniel Zohary, The Hebrew University of Jerusalem, Jerusalem, Israel

CHAPTER 3: DESERTS AND THE PLANET — LINKAGES BETWEEN DESERTS AND NON-DESERTS

Lead author:

Uriel Safriel, The Hebrew University of Jerusalem, Israel

Contributing authors:

Pinchas Alpert, Tel-Aviv University, Tel-Aviv, Israel

Uzi Avner, Ben-Gurion University of the Negev, Eilat, Israel

Yoram Ayal, Ben-Gurion University of the Negev, Midreshet Ben-Gurion, Israel

Niels H. Batjes, ISRIC - World Soil Information, Wageningen, The Netherlands

Noah Brosch, Tel Aviv University, Tel Aviv, Israel

Rex Cates, Brigham Young University, Provo, Utah, USA

Dan Cohen, The Hebrew University of Jerusalem, Jerusalem, Israel

Thomas Demére, San Diego Natural History Museum, San Diego, California, USA

David Dent, ISRIC - World Soil Information, Wageningen, The Netherlands

Exequiel Ezcurra, San Diego Natural History Museum, San Diego, California, USA

Avi Golan-Goldhirsh, Ben-Gurion University of the Negev, Midreshet Ben-Gurion, Israel

Elli Groner, Ben-Gurion University of the Negev, Midreshet Ben-Gurion, Israel

Emanuel Mazor, The Weizmann Institute of Science, Rehovot, Israel

Scott Morrison, The Nature Conservancy, California, USA

Christian Nellemann, UNEP GRID-Arendal /NINA, Lillehammer, Norway

Berry Pinshow, Ben-Gurion University of the Negev, Midreshet Ben-Gurion, Israel

Boris A. Portnov, University of Haifa, Haifa, Israel

Daniel Rosenfeld, The Hebrew University of Jerusalem, Jerusalem, Israel

Haim Shafir, Tel-Aviv University, Tel-Aviv, Israel

William H. Schlesinger, Duke University, Durham, North Carolina, USA

Moshe Schwartz, Ben-Gurion University of the Negev, Midreshet Ben-Gurion, Israel

Ina Tegen, Leibniz Institute for Tropospheric Research, Leipzig, Germany

Anastasios Tsonis, University of Wisconsin-Milwaukee, Wisconsin, USA

Andrew Warren, University College London, London, UK

Robin P. White, U.S. Geological Survey, Kearneysville, West Virginia, USA

Box author:

Uriel Safriel, The Hebrew University of Jerusalem, Jerusalem, Israel

CHAPTER 4. STATE AND TRENDS OF THE WORLD DESERTS

Lead authors:

Stella Navone, Universidad de Buenos Aires, Buenos Aires, Argentina

Elena Abraham, Instituto Argentino de Investigaciones de las Zonas Áridas, Mendoza, Argentina

Contributing authors:

Marta Bargiela, Universidad de Buenos Aires, Buenos Aires, Argentina

David Dent, ISRIC • World Soil Information, Wageningen, The Netherlands

Cecilia Espoz-Alsina, Pontius S.A., Rochester, New York, USA

Alejandro Maggi, Universidad de Buenos Aires, Buenos Aires, Argentina

Elma Montaña, Instituto Argentino de Investigaciones de las Zonas Áridas, Mendoza, Argentina

Scott Morrison, The Nature Conservancy, California, USA

Gabriela Pastor, Universidad de Buenos Aires, Buenos Aires, Argentina

Hector Rosatto, Universidad de Buenos Aires, Buenos Aires, Argentina

Mario Salomón, Instituto Argentino de Investigaciones de las Zonas Áridas, Mendoza, Argentina

Dario Soria, Instituto Argentino de Investigaciones de las Zonas Áridas, Mendoza, Argentina

Laura Torres, Instituto Argentino de Investigaciones de las Zonas Áridas, Mendoza, Argentina

Box authors:

Elena Abraham, Instituto Argentino de Investigaciones de las Zonas Áridas, Mendoza, Argentina

Marta Bargiela, Universidad de Buenos Aires, Buenos Aires, Argentina

Walter Massad, Universidad de Buenos Aires, Buenos Aires, Argentina

Clara Movia, Universidad de Buenos Aires, Buenos Aires, Argentina

Stella Navone, Universidad de Buenos Aires, Buenos Aires, Argentina

Fidel A. Roig, Instituto Argentino de Investigaciones de las Zonas Áridas, Mendoza, Argentina

Sergio Roig-Juñent, Instituto Argentino de Investigaciones de las Zonas Áridas, Mendoza, Argentina

Laura Torres, Instituto Argentino de Investigaciones de las Zonas Áridas, Mendoza, Argentina

CHAPTER 5. CHALLENGES AND OPPORTUNITIES: CHANGE, DEVELOPMENT, AND CONSERVATION

Lead author:

Andrew Warren, University College London, London, UK

Contributing authors:

Martin A. Green, University of New South Wales, Sydney, Australia

Stefanie M. Herrmann, University of Arizona, Tucson, Arizona, USA

Conrad Roedern, Solar Age Namibia Ltd., Windhoek, Namibia

Uriel Safriel, The Hebrew University of Jerusalem, Israel

CHAPTER 6. DESERT OUTLOOK AND OPTIONS FOR ACTION

Lead authors:

Stefanie M. Herrmann, University of Arizona, Tucson, Arizona, USA

Charles F. Hutchinson, University of Arizona, Tucson, Arizona, USA

Section authors:

David Dent, ISRIC - World Soil Information, Wageningen, The Netherlands

Scott Morrison, The Nature Conservancy, California, USA

Kakuko Nagatani, DEWA, UNEP-ROLAC, Mexico City, Mexico

Christian Nellemann, UNEP GRID-Arendal /NINA, Lillehammer, Norway

Andrew Warren, University College London, London, UK

Contributing authors:

Abiliti Abdukadir, Xinjiang Institute of Ecology and Geography, Urumqi, Xinjiang, China.

Hugo Ahlenius, UNEP GRID-Arendal, Lillehammer, Norway

Robert Alkemade, Netherlands Environmental Assessment Agency (MNP), Bilthoven, The Netherlands

Michel Bakkenes, Netherlands Environmental Assessment Agency (MNP), Bilthoven, The Netherlands

Ben ten Brink, Netherlands Environmental Assessment Agency (MNP), Bilthoven, The Netherlands

Exequiel Ezcurra, San Diego Natural History Museum, San Diego, California, USA

Ole-Gunnar Støen, Norwegian University of Life Sciences, Aas, Norway

Bjørn P. Kaltenborn, Norwegian Institute for Nature Research, Lillehammer, Norway

Lera Miles, UNEP World Conservation Monitoring Centre, Cambridge, UK

Tonnie Tekelenburg Netherlands Environmental Assessment Agency (MNP), Bilthoven, The Netherlands

Mary Seely, Desert Research Foundation of Namibia, Windhoek, Namibia

Ingunn Vistnes, Norwegian University of Life Sciences, Aas, Norway

Andrew Warren, University College London, London, UK

Participants of the Review Meetings in Mendoza, Argentina (6-8 September 2005) and Gobabeb, Namibia (24-27 January 2006):

Elena Abraham, Instituto Argentino de Investigaciones de las Zonas Áridas, Mendoza, Argentina

David Dent, ISRIC - World Soil Information, Wageningen, The Netherlands

Exequiel Ezcurra, San Diego Natural History Museum, San Diego, California, USA

Stefanie M. Herrmann, University of Arizona, Tucson, Arizona, USA

Timo Maukonen, UNEP DEWA, Nairobi, Kenya

Scott Morrison, The Nature Conservancy, California, USA

Kakuko Nagatani, DEWA, UNEP-ROLAC, Mexico City, Mexico

Stella Navone, Universidad de Buenos Aires, Buenos Aires, Argentina

Christian Nellemann, UNEP GRID-Arendal /NINA, Lillehammer, Norway

Uriel Safriel, The Hebrew University of Jerusalem, Israel

Mary Seely, Desert Research Foundation of Namibia, Windhoek, Namibia

Ashbindu Singh, UNEP DEWA - North America, Washington, D.C., USA

Guido Soto, Water Center for Arid Zone and Semi Arid Zone in Latin America and the Caribbean, La Serena, Chile

Wang Tao, Cold and Arid Regions Environmental and Engineering Research Institute, Chinese Academy of Sciences, Lanzhou, China

Andrew Warren, University College London, London, UK

Reviewers:

H. Gyde Lund, Forest Information Services, Gainesville, VA, USA

Mohamed Sessay, UNEP DGEF, Nairobi, Kenya

Mark Stafford-Smith, Desert Knowledge Cooperative Research Centre, Alice Springs, Australia

Anna Tengberg, UNEP DGEF, Nairobi, Kenya

Wu Zhongze, National Bureau to Combat Desertification (NBCD), Beijing, China

UNEP technical staff:

Marcus Lee, UNEP DEWA, Nairobi, Kenya

Timo Maukonen, UNEP DEWA, Nairobi, Kenya

Kakuko Nagatani, DEWA, UNEP-ROLAC, Mexico City, Mexico

Bruce W. Pengra, USGS Center for Earth Resources Observation and Science, Sioux Falls, South Dakota, USA

Ashbindu Singh, DEWA, UNEP-North America, Washington D.C., USA

Hua Shi, UNEP/GRID, Sioux Falls, South Dakota, USA

Support staff at UNEP Nairobi and Mexico City:

María Teresa Hurtado Badiola

Beth Ingraham

Ruth Mukundi

Jennifer Odallo

Esther Mendoza Ramos

Audrey Ringler

Global Deserts Outlook Collaborating Centres

UNEP would like to acknowledge the institutional support from the following Collaborating Centres, without which this report would not have been possible.

- San Diego Museum of Natural History, United States

- Instituto Argentino de Investigaciones de las Zonas Áridas (IADIZA), Argentina

- Gobabeb Training and Research Centre, Namibia

Acknowledgements

The Editor and the Lead Authors want to acknowledge the continuous help and support received throughout this project from UNEP's offices. We want to thank all the technical staff from UNEP-ROLAC in Mexico City, and UNEP/GRID in Sioux Falls. Very especial thanks to Kakuko Nagatani, Ashbindu Singh, Hua Shi, and Bruce W. Pengra, who put their outstanding talent and their dedication fully behind this project, and made the production of this volume possible in just seven months of very hard work. We also want to thank the two project liaisons at UNEP's Division of Early Warning and Assessment (DEWA) in Nairobi, Timo Maukonen and Marcus Lee. Their technical knowledge — not to mention patience with the group, diplomatic skills, and good-humoured attitude — made our work a true pleasure. Finally, our thanks to all the support staff at UNEP Nairobi and Mexico City: Esther Mendoza Ramos, María Teresa Hurtado Badiola, Beth Ingraham, Jennifer Odallo, and Audrey Ringler, as well as John Mugwe who organized the WIKI site and responded quickly and patiently to all our requests and queries. We are deeply obliged to all of them.

Mary Seely, lead author of chapter 2, wishes to thank the invaluable assistance received from Inge Henschel, Janis Klimowicz, Judith Lancaster, and Lisa Wable. Richard Felger thanks the Wallace Research Foundation for their generous support of the research projects that allowed the writing of the box on Sonoran Desert plant uses. Sabine Schmidt wishes to acknowledge, in the framework of Mongolian-German Technical Cooperation, the projects "Nature Conservation and Buffer Zone Development" (1995-2002), and "Conservation and Sustainable Management of Natural Resources – Gobi Component" (GTZ, German Technical Cooperation), implemented by IPECON – Initiative for People Centred Conservation of NZNI (New Zealand Nature Institute), who facilitated participatory planning with herder communities and piloted people-centred approaches to conservation in the Gobi. Uriel Safriel, lead author of chapter 3, acknowledges the help received from Meir Pener from the Hebrew University of Jerusalem for providing material and advice for the locust section; to Zohara Yaniv from the Agricultural Research Center, Bet-Dagan, Israel, and Ofer Bar-Yosef from Harvard University for advice regarding the medicinal plants and archaeological sections, respectively; to Yossi Leshem from Tel Aviv University and Peter Berthold from Max-Planck-Institut für Ornithologie, Vogelwarte Radolfzell for allowing the use of the map and their information on the migration of the white stork; and to William Schlesinger from Duke University for providing insightful comments on the chapter. Scott Morrison thanks Brian Cohen, Kirk Klausmeyer, David Mehlman, Bob McCready, and Mark Reynolds for contributions to and review of earlier versions of the bird migration section of chapter 3, as well as Jonathan Hoekstra and the entire Global Habitat Assessment Team of The Nature Conservancy for support in producing the book's Appendices. Elena Abraham and Stella Navone, lead authors of chapter 4, would like to thank Nelly Horak for her skillful help with the translation of their manuscript. Stefanie Herrmann and Charles Hutchinson, lead authors of chapter 6, wish to thank Frank Eckardt, from the University of Cape Town, for assistance in the processing of the vegetated sand dune images.

Lastly, the Editor wants to thank the warm hospitality provided at our two workshop meetings by the wonderful staff of *Instituto Argentino de Investigaciones de las Zonas Áridas* (IADIZA) in Mendoza, Argentina, and that of the Gobabeb Training and Research Centre in Gobabeb, Namibia. The stunning natural beauty of the two places and the incredible research that is being done at both institutions was a source of inspiration for all of us.

Financial support for the Global Deserts Outlook was provided by the counterpart contribution from the Government of Sweden, and by the Partnership Programme between UNEP and the Netherlands Minister for Development Cooperation.

This publication is printed on chlorine and acid-free paper from sustainable forests.